Towards Humane Technologies

TRANSDISCIPLINARY STUDIES
Volume 2

Scope
Transdisciplinary Studies is an internationally oriented book series created to generate new theories and practices to extricate transdisciplinary learning and research from the confining discourses of traditional disciplinarities. Within transdisciplinary domains, this series publishes empirically grounded, theoretically sound work that seeks to identify and solve global problems that conventional disciplinary perspectives cannot capture. Transdisciplinary Studies seeks to accentuate those aspects of scholarly research which cut across today's learned disciplines in an effort to define the new axiologies and forms of praxis that are transforming contemporary learning. This series intends to promote a new appreciation for transdisciplinary research to audiences that are seeking ways of understanding complex, global problems that many now realize disciplinary perspectives cannot fully address. Teachers, scholars, policy makers, educators and researchers working to address issues in technology studies, education, public finance, discourse studies, professional ethics, political analysis, learning, ecological systems, modern medicine, and other fields clearly are ready to begin investing in transdisciplinary models of research. It is those many different audiences in these diverse fields that we hope to reach, not merely with topical research, but also through considering new epistemic and ontological foundations for of transdisciplinary research. We hope this series will exemplify the global transformations of education and learning across the disciplines for years to come.

Towards Humane Technologies
Biotechnology, New Media and Ethics

Naomi Sunderland
Queensland University of Technology

Phil Graham
Queensland University of Technology

Peter Isaacs
Queensland University of Technology

Bernard McKenna
University of Queensland

SENSE PUBLISHERS
ROTTERDAM / TAIPEI

A C.I.P. record for this book is available from the Library of Congress.

ISBN 978-90-8790-444-9 (paperback)
ISBN 978-90-8790-445-6 (hardback)
ISBN 978-90-8790-446-3 (e-book)

Published by: Sense Publishers,
P.O. Box 21858, 3001 AW
Rotterdam, The Netherlands

Printed on acid-free paper

TABLE OF CONTENTS

ACKNOWLEDGEMENTS

The editors would like to acknowledge all participants in the Towards Humane Technologies Conference held at the University of Queensland, Ipswich, Australia in July 2002. The papers in this volume were inspired by that conference. We would also like to acknowledge all sponsors for the event including Biotechnology Australia, the Queensland University of Technology Program in Applied Ethics and Human Rights, the University of Queensland Business School, and the Queensland Department of State Development.

SECTION I INTRODUCTION

JEREMY HUNSINGER

SERIES INTRODUCTION

We chose transdisciplinary studies as the basis for our book series because it embodies a universally embracing set of practices and knowledges. Transdisciplinary research comprises an approach to research that recognizes that for many research questions, disciplinary perspectives cannot provide a complete or even an adequate understanding of the systems and processes involved. Recourse to transdisciplinary modes of inquiry often occurs upon the recognition that the disciplinary or inter-disciplinary perspectives used comprehend the object of research are insufficient, and that no addition of another disciplinary or interdisciplinary technique will be sufficient. The moment that a research team recognizes the insufficiency of their approach based on an understanding of the globality of the research object, they can either forgo further research or turn to the broader dialogues of transdisciplinarity (Genosko, 2002). The research team can engage with other disciplines that will lead to meta-questions regarding axiomatics and axiologies (Hunsinger, 2005).

A discipline's axiomatics and axiologies inhabit its community of researchers and. as such, defines its territory, perspectives and boundaries. Once a research team takes up questions of axiology and axiomatics in relation to objects of research, in order to build a global understanding of the object of research, then their research is transdisciplinary research. Accordingly, the research team will depend on many different knowledges and understandings that will have to function in dialogues with each other, as the researchers continually explore the relations and relationships possible between the various contending knowledges. These dialogues, allowing for new interpretations of central questions, help to dissolve the disciplinary boundaries that constrain perspectives in disciplinary research. This dissolution opens up the axiologies and axioms to reconstruction, which reconstructs the horizon of understanding that was built into the disciplinary or interdisciplinary perspectives. The opening of the horizon of understanding opens new questions for inquiry and research.

In comparison to interdisciplinary research, which is bound up with the contestation and borrowing of techniques from various disciplines in order to map a knowledge domain, transdisciplinary research questions the very nature of the map, and the foundations of the techniques that produced it. Transdisciplinary research cuts across two or more disciplines, unifying them at the foundations through constructive dialogues and practices that lead to shared assumptions and thus a more global understanding. Interdisciplinary research, much like disciplinary research can participate in transdisciplinary research, but the terms are not synonyms nor are their modes of praxis similar. Central to interdisciplinary research is the

Naomi Sunderland et al. (eds.), Towards Humane Technologies: Biotechnology, New Media and Ethics, 3–6.

idea that knowledge can be shared, and techniques and insights can be transposed or relocalized, across knowledge domains, whereas transdisciplinary research takes as it central idea that the practices of production of knowledge are bound to axiomatic and axiological assumptions that constrain the sharing of knowledge, and that sharing of knowledge from a disciplinary or interdisciplinary perspective is to share a perspective on the knowledge that then has to be transposed or translated across knowledge domains. Transdisciplinary research engages in a negotiation and translation of the axiomatics and axiologies that construct the perspective, hoping to reconstruct those assumptions into a new shared whole that then allows greater understanding about the object than is possible from any single perspective. By engaging in dialogue at the level of the foundations of our knowledges and research practices, transdisciplinary research find ways to resolve the contestations of the interdisciplinary and disciplinary translations and trans-positions. In resolving these contestations over boundaries through discussion of foundations, transdisciplinary research engages in a reconstruction of understanding that can engage broader publics, across disciplines and across divided ethos (Snow, 1993).

Disciplinary, interdisciplinary, and transdisciplinary research each contribute to the system of knowledge production in our current age of late capitalism. Each produces results that have merits for society, and each has unique relationships to the way it understands the systems of knowledge and the production of knowledge in relation to those systems. That is not to say that the knowledge produced from these modes of research cannot have uniquely different ends. Instead, the ends of research depend much more on the context of production of knowledge than on the mode of research. The context of production of knowledge stands in relation to the broader context of the audience of the research, the public's understanding of the research, and its applicability to the general mode of production found in our current world.

It would be a cliché to say that research is changing, or scholarly production is changing, just as most production is changing in late capitalism. All are clichés. Inasmuch as these clichés are games, in Lyotard's sense, the game/clichés are producers of systems of legitimation that concretize the changes as they happen. They are not part of systematic metanarratives of progress, but operate merely as the everyday practices of research, scholarly production and production in general in the world. There is constitutive difference in our thoughts about knowledge that cannot be negotiated without some of these cliches. When we talk about research, we say things about the practice that approximate "This is how research is done," and then we abstractly represent the knowledges about those practices, "These are our beliefs about research." The movement from practice to belief and fact is one of textual transformation in relation to a broad field of signs existing in a system of legitimation and justification (Thévenot, 1984, 2002; Latour, 1986; Boltanski and Thévenot, 2006). The cliché of change in relation of research help perform the paralogics necessary to move between our practices and our knowledges (Lyotard, 1984).

The legitimation of transdisciplinary research is often found precisely in its engagement with the popular, the political, and the applied research that is frequently missed by disciplinary and interdisciplinary efforts. Transdisciplinary research is often driven by an interest in solving a particular problem that is too complex to have been successfully dealt with from any given disciplinary perspective (Gibbons, et. al, 1994). Its legitimation is found in the applicability and thus fundability of its research regimes, which ends up being a pragmatic aspect as much as a defining condition, not in its attempt to resolve axiological or axiomatic conflicts.

The new generation of transdisciplinary researchers and the problems they address are, on one level, working collectively on a plurality of problems. However, it should be noted that there is no necessary unity to the problems they approach. Other than a central concern with the dissensus of knowledge production and knowledge systems bound by disciplines and interdisciplinary research, these researchers are working individually or in teams pursuing their own particular resolutions to the complex issues that they are engaged with. The plurality of modes of research in which they are engaged is representative of the recombinative possibilities of questions and perspectives that they and their teams engage in pursuing.

Researchers researching research can admit that the existence of a plurality of modes of research and the different ways of producing the wide variety of knowledges about the world demonstrates the transformation of research. As we move away from our unitary and disciplinary conceptions of knowledges toward questions that can only be answered from transdisciplinary inquiry, we are faced with the problems central to opening any new dialogue constituted in the contestations of in the foundational axiomatics and axiologies of what we think of knowledge itself (Hunsinger, 2008). Even our ability to communicate the nature of our individual knowledge from one discipline to another becomes a seemingly insurmountable organizational problem. The reconstruction of research and research communities has always been a social and organizational problem but, as research is continually changing, it is becoming more transdisciplinary.

It is not surprising that more people are challenging the meta-problems of pluri-disciplinarity, such as those found in transdisciplinarity. Our book series, Transdisciplinary Studies, aims to challenge those problems as we search for texts that dissolve disciplinary boundaries, build new axiomatic and axiological relationships, and provide pathways for future research.

Towards Human Technologies, the first in the series, opens a discussion on three interrelated topics. Starting by questioning the interrelationships of biotechnology, new media and ethics, the editors juxtapose a series of articles from significant scholars centered around contextualizing and contesting the knowledges of biotechnologies and media, their interrelationships, and, in the end, provide a basis for transdisciplinary dialogues on these topics.

REFERENCES

Boltanski, L., & Thévenot, L. (2006). *On justification: Economies of worth (Princeton Studies in Cultural Sociology)*. Princeton University Press.

Genosko, G. (2002). Felix Guattari: *An aberrant introduction*. London: Continuum.

Gibbons, M., Limoges, C., Nowotny, H., Schwartzman, S., Scott, P., & Trow, M. (1994). *The new production of knowledge: The dynamics of science and research in contemporary societies*. Thousand Oaks, CA: Sage Publications Ltd.

Hunsinger, J. (2005). Toward a transdisciplinary internet research. *The Information Society, 21(*4), 277–279.

Hunsinger, J. (2008). The virtual and virtuality: Towards dialogues of transdisciplinarity. In N. Panteli & M. Chiasson (Eds.), *Exploring virtuality within and beyond organizations*. London: Palgrave Macmillan.

Latour, B., & Woolgar, S. (1986). *Laboratory life*. Princeton University Press.

Lyotard, J. F. (1984). The postmodern condition: A report on knowledge. In *Theory and history of literature* (Vol. 10). University of Minnesota Press.

Nowotny, H., Scott, P., & Gibbons, M. (2001). *Re-thinking science: Knowledge and the public in an age of uncertainty*. London: Polity Press.

Snow, C. P. (1993). *The two cultures*. Cambridge University Press.

Thévenot, L. (1984). Rules and implements: Investment in forms. *Social Science Information, 23*(1), 1–45.

Thévenot, L. (2002). Conventions of co-ordination and the framing of uncertainty. In E. Fullbrook (Ed.), *Intersubjectivity in economic*s (pp. 181–197). New York, NY: Routledge.

BERNARD MCKENNA

1. INTRODUCTION

Given the considerable space used up in writing about biotechnology, one may well ask why this book has been written. The unique perspective that this book collectively provides is that it links biotechnology with media and citizenship. Biotechnology here means the direct manipulation of genetic material, and not the broader conceptualisation of any manipulation of organic life such as food fermentation. Media has a dual meaning: the representation of biotechnology in the (mass) media and the important role of data storage and manipulation in the biotechnological processes. Citizenship is concerned with the moral responsibility for the wellbeing of society now and into the future. The ethical disposition of the book is to provide spaces for alternative and dissident voices so that theorists and policy makers have the opportunity to step outside the hype that surrounds biotechnology and consider the issues from a position of humanity.

The book is in three sections:
i.   Contextualising Biotechnology and New Media;
ii.   Sites of Contestation; and
iii.   Responses.

In the first section, Contextualising Biotechnology and New Media, Phil Graham begins by providing a political economic framework for considering biotechnology. Eugene Thacker considers how our extropian tendencies lead to the potential de-humanising of the body. Ross Barnard and Damian Hine continue in this line by considering how positivist tendencies in biotechnological education need to be tempered by more humane concerns. Joseph Vogel considers the impact of commodified biotechnology of the North from the persepctive of the megadiverse South. The Sites of Contestation considered in the second section raises critical issues of concern. Bonfiglioli considers the way in which the media construct our perceptions of biotechnology. Clapton expresses concern that the marginalizing of certain groups of people could lead to inhumane choices that deal with procreation, foetal diagnosis, and disability. Milligan sees in the discourses of ethical institutional frameworks notions of individual choice, individualism, and rationality. Lassen looks more closely at the discourses of biotechnological to traces of the need for progress, economic benefit, and legal-technical issues, which empirically reinforces the concerns raised by other contributors. The book's closing section, Responses, provides useful directions for future ethics-based considerations of biotechnology, particularly from a discourse perspective. Peter Isaacs stresses the need to understand

Naomi Sunderland et al. (eds.), Towards Humane Technologies: Biotechnology,
New Media and Ethics, 7–14.

our embeddedness by critically evaluating the discourses that valorise reason, autonomy, and freedom by listening to alternative voices outside the established traditions and norms. Clare Christensen's analysis of what it means to be scientifically literate also emphasizes that the capacity to step outside our social constructedness is crucial to effective citizenship when dealing with biotechnological questions. Finally, Naomi Sunderland explains how the processes of mediation within biotechnology industry and research and in the broader mediation to society can detach or sideline our humane impulse thereby producing social, political, and economic outcomes that we would not have intended to happen.

Implicit throughout the book is the understanding that biotechnology is not just concerned with a particular branch of applied science. Because biotechnology in the current phase of capitalism is commodified, it is woven into the political and economic fabric of society. Using a Marxist approach, Phil Graham's political-economic analysis considers how social phenomena, material and immaterial, are valued, distributed, and exchanged, while also considering how the collective and individual interests are settled. Essentially he tracks two tributaries leading to the current state of biotechnology: the link between biology and engineering, and the emergence of corporate capitalism. Up to the 1960s, he argues, biotechnology was predicated on engineering people to fit machinic purposes. However, a cultural turn including a greater concern with the state of the environment, led to a totalizing notion of biotechnology controlling and automating everything. Then, pointing out that contemporary corporatist capitalism relies heavily on perceptions of future value, Graham says that biotechnology is particularly suited for such a phase. While companies such as Biogen have produced worthwhile biotechnological products, the whole biotechnological industry is essentially concerned with the potential for investors to become wealthy by buying stock early. With so much speculative investment resting on the outcomes, the potential for risk will be understated and the potential for benefit will be overstated, he says, and problems will be shaped in such a way that favors corporate capitalism. Understanding this broader political economic context should lead us to re-consider the human issues involved and to set aside the economic imperatives. After all, biotechnology is very much concerned with the essence of life itself.

The relationship between the technologies of data-collection and the technologies of human biology has not been considered much outside of speculative fiction. This is why Eugene Thacker's paper is a timely effort to set this right. Designating the relationship between information technologies and biotechnologies as 'biomedia', he then places this phenomenon in the broader sociocultural context of posthumanism, which has two opposing strands: extropianism and a critical perspective. Because extropianism is technophilic, it is relatively sanguine about the implications of the changing technologies; however, the more critical perspective raises quesions about what it now means to have a body, to be a body, to create a body. Because extropian discourses incorporate Enlightenment discursive concepts such as progress, optimism, open society, and rational thinking – Thacker calls it smuggling humanist-based conceit – opposition can appear reactionary, a point made also by Sunderland. At the heart of his concern is the blithe ontological separation between human and machine that the extropians make, which paves the way to neutralize technology. The more critical posthumanists acknowledge the advances made by biotechnology, but are less prone to utopic considerations and are more interrogative of the potential implications of the 'new relationships between human and machine, biology and technology, genetic and computer information'. Drawing on the work of Claude Shannon and Norbert Wiener, Thacker concludes that when any sort of information is transmitted, it becomes disconnected from the language, culture, and context of real human life. The implication of this for biotechnology is that when the body is reconstrued as information, this disconnection occurs, rendering it easier for the body to be technically manipulated, controlled, and monitored. This happens through a process of translating, recoding, and decoding. In a sense, Thacker's concern is the ethical implications of a biotechnology that detaches human life and life components (e.g., stem cells) from humanity – 'lateral transcendence' – by the processes of biomedia. Such possibilities are not being considered in the headlong rush by First World countries to achieve bio-technological breakthroughs.

Barnard and Hine begin their paper by acknowledging the importance of traditional scientific methods in developing effective modern medicine that has helped to prevent disease, reduce childbirth risk, and increase our life expectancy. Nonetheless, they argue, the positivist foundations of modern science and Cartesian dualism must now be tempered by contemporary ontological and epistemological challenges. Furthermore, they argue, because biotechnology is based on a utilitarian ethic that is product focussed, the crucial role of fundamental research is being forsaken. In the Australian context, they point out that the increasing market orientation of universities is reducing science funding, especially in pure science, and is reducing the quality of the educational outcome. Collectively these factors reinforce positivist and Cartesian science. They want to replace this with a new science education paradigm that maintains technical and practical competence, but presents students with an understanding of the epistemological assumptions that underlie any body of knowledge. Indeed, students should be capable of deconstructing their knowledge base. Such a capability is important, they say, for the sorts of dialogues in which future scientists should engage, dialogues that incorporate the social and the ethical as well as the scientific.

As with any 'commodity', biological products have been commodified by capitalism. Joseph Vogel applies some basic economic tools to consider bio-prosepcting. He draws our attention to the fact that the mostly Southern megadiverse nations that provide the raw materials receive a 'picayune', and may, in fact, receive nothing when such materials are declared to be the property of no one. By contrast, the Northern nations that use these materials in biotechnological products are provided with monopoly patents, particularly in the U.S. which enforces the Trade Related Intellectual Property Rights (TRIPS) conventions on other countries. That is, the North's *Biodiscovery* is the South's *Biofraud.*

The media's strong influence over the public perception of biotechnology, particularly gene technology motivates Catriona Bonfilioli's chapter. Bonfiglioli points out that, despite claims to the contrary, positive coverage of biotechnological issues dominates negative coverage. In particular, she argues that genetic determinism is embedded in mass media texts. The effect is to increase public acceptance of many genetic technologies by normalizing genetic explanations for human problems. As a result, she fears that medical technologies can pave the way for genetic determinism. Given this intense focus on the genetic aspect, environmental components of disease will be rendered marginal or irrelevant. She refers to the common locution 'the gene for' as typical of such distortions, because it confuses genetic markers for genes. Although well-intentioned and useful, The Human Genome Project intensifies this focus on genes as the control centre of human biology. These epigenetic developments run the risk of being appropriated by interests such as the insurance industry attempting to minimise risk. As well, pre-natal testing for 'defects' can lead us unthinkingly towards a new eugenics, but wrapped in a discourse of choice and rights. Thus, we need to be vigilant in how we 'frame' genetic research, policy, and public reporting. Unintended consequences such as eugenics and the diversion of resources and funds away from other areas of research and activity that also have legitimate claims need to be guarded against.

The discourse of biotechnology, which is infused with future-oriented intimations of hope –a characteristic identified also by Graham – based on cures and controls for disease and disability, implies an enhanced society. However, Jayne Clapton draws our attention to the citizenship implications of a society that necessarily eliminates those 'things' or beings deemed unworthy of life. Using the word-play of re-membering as a process of putting back together, as it were, as well as recollection, Clapton puts the case that contemporary imperatives should be scrutinized for what might be marginalized or excluded in our deliberations. Ethical practice, she argues, requires reflection based on *anamnesis* of past practices, not the comfortable process of amnesia. By denying access to groups of people who do not have access to ethical decision making, by not incorporating their voices, and by sanitizing practices and sites, we are complicit in amnesic practices, she says. Thus, contestation is muted or non-existent to the advantage of dominant discourses and dominant social groups. Because biotechnology frames the world primarily from a strongly reductionist and analytical philosophical paradigm, it necessarily creates a tension with a humanistic paradigm. Consequently, the biotechnological orientation tends to represent the person as 'micro parts', rather than a being within an expansive metaphysical realm. Associated with the scientific paradigm are other vast discourses of law and politics. Although proponents of biotechnology do not promote eliminating certain types of disabled people, there is nonetheless the risk that interpreting people's lives in particular ways will orient choices about procreation, foetal diagnosis, and disability in particular ways. For example, however well-meaning a decision-maker may be, if they perceive disabled people as dependent and non-productive, then they are unlikely to be positively valued.

When we consider the history of the 'institution' or 'asylum', it should be obvious, Clapton argues, that practices and discourses developed around notions of care and control towards anomalous Others can lead to undesirable outcomes. Similar notions used in contemporary discussions about preventing the life of a person who might 'suffer' is outdated she says because the suffering is often the result of the different relationships between 'normal' people and the 'other'. Incorporating narratives of the disabled as not othered will provide some countervailing force to the powerful rational processes of biotechnology. Thus tropes of acceptance, mutuality and interdependence will counter those of suffering, limitation and loss, thereby altering the balance of likely outcomes. Whether the powerful disciplines of contemporary biotechnological practices admit such narratives and tropes affects the ethics, practices and possibility of life itself.

The tension between liberating and eugenic tendencies in the new reproductive technologies is Eleanor Milligan's concern. This tension has been made particularly relevant given the widespread use of prenatal screening (up to 90% in First World countries). The apparent resolution of this tension is the notion of a patient's informed consent. However, Milligan identifies several concerns with this concept particularly in the context of the entrenched institutional pathways of routine screening, counselling, and predefined outcomes (over 85% of foetuses prenatally diagnosed as abnormal are aborted). As Milligan points out, the availability of this knowledge creates an expectation of some action, but it also raises the problem of defining what is abnormal or a disability. Of concern is that the decision-making process in this complex area occurs within a discourse of choice, and of uncontested science. Basing such life and death judgments on 'choice', itself part of a larger discourse of individualism, does not guarantee ethical integrity, she points out. In any case, the concept of informed consent to engage in testing and subsequent actions is also highly questionable for a number of reasons. A purely rational approach does not allow for non-rational aspects such as intuition and humane impulses. As well, the decision maker is not necessarily cognizant of the 'relevant' knowledge needed, and the available options are predetermined. Milligan points out that the predictive capacity of reproductive technologies is highly questionable in that pre-natal tests suggesting potential abnormality frequently are not borne out. An ensuing further ethical dilemma is created by the fact that the capacity to treat has not kept pace with the capacity to diagnose, and this then creates further imperatives for more technological interventions in the whole natal process. As a consequence, Milligan argues for an expanded ethical framework beyond the current formulaic procedure rooted in notions of individual choice, individualism, and reason-based science. Consequently, this new ethics must be seen as involved in human relationships beyond just the individual. It must also be based on an understanding of the interconnectedness of biotechnology as a science and as a political, economic, and social practice.

The possibility of dialogue between proponents and sceptics of biotechnology is Inger Lassen's concern. She linguistically analyses journalistic interviews of scientists, politicians, and a consumer organization representative in Denmark on the topic of food biotechnology. Infusing the discourse of these subjects were the intertextual traces of safety, the need for progress, concern for less well-fed people, economic beneficiaries, and legal-technical issues of patents. Rather than seeing these as points of difference, the implication is that these intertextual elements might form the basis of dialogue. A further optimistic element of the research is that both sides seemed to share the same goal, a world without suffering.

Peter Isaacs provides a genealogy of the dominant contemporary view that most biotechnological developments are either ethical or not matters for ethical consideration, thereby marginalizing or silencing criticism. He does this by providing an overview of the 'technological impulse' and a brief history of the Western ethical tradition. By technological impulse, Isaacs means that technology has increasingly dominated culture through the three cultural stages of hunter gatherer, agrarian, and the scientific-technological. However, the goal has changed from the most 'primitive' instinct of survival to the current situation in which the Industrial stage is giving way to the IT and Biotechnological Revolutions. Furthermore, the focus of activity has shifted from the natural environment to that of human meanings (information, knowledge, entertainment, media, culture) and to the environment that constitutes humans themselves as embodied, living organisms. Accompanying this has been the political-economic shift from industrial to consumer capitalism, the major characteristics of which are entrepreneurialism and commodification, even of bodies. The ethical tradition has also been complex, moving from polytheism to secular postmodernism. Isaacs locates Kant as the primogenitive source of contemporary secularism because of his assumption that individual decision-making constituted the moral form of life (of course, individual conscience was given considerable impetus by the Protestant revolt). God as the source of goodness was replaced by the categorical imperative and the assumption that, in the social order, people act rationally when they comply with the universal moral law. Concomitant with this, Isaacs claims, is that, in the natural order, people act rationally when they comply with scientific laws.

With technological development in consumer goods, transport, medicine, and luxury items, 'paradise' for many moved from the hereafter to the present, and so the meaning and purpose of life moved from the transcendental to the immanent. Consistent with this have been the 'conditions of interiority' that valorise not just reason, but also personal autonomy and freedom. All of this may seem to provide little encouragement to those who seek to challenge the hegemonic suite of discourses that incorporate achievement, personal freedom, progress, and entrepreneurial capitalism. However, Isaacs suggests that one way forward is to listen to voices outside the established traditions if we wish to reappraise what it is to be human, and what constitutes the good life. To be truly critically reflective, we must also negotiate our 'embeddedness as human beings'. By becoming aware of our own biographies, understandings, hopes and fears at a personal level, and our embeddedness in culture, institutions, and communities at the social level, we are more likely to become aware of the assumptions that underlie our practical and ethical engagement with biotechnology.

The fact that only 5% and 15% of the adult population in most Western countries are scientifically literate motivates Clare Christensen to consider what scientific literacy is. This is particularly important given that our 'knowledge' society is primarily scientific and technological, and that our 'risk' society is concerned with applying new technologies. Effective citizenship requires that people have the capacity to deal with the ethical, social, and physical implications of technolgical development and change. However, defining scientific literacy is difficult. For example, in broad terms, there are different emphases between Anglophone and European approaches to scientific literacy. The former see literacy as the ability to cope with scientific information meaningfully, as well as being interested and confident. The 'European' approach, however, is more concerned with developing a 'scientific culture'. The OECD seems to provide useful guidance by drawing together these approaches such that schools would develop 'the capacity to use scientific knowledge, to identify questions and to draw evidence-based conclusions in order to understand and help make decisions about the natural world and the changes made to it through human activity'. In the face of exponentially growing knowledge and the technological imperative, it appears that relying on providing only a knowledge base at school, though necessary, will not in itself provide the requisite skills of effective citizenship in dealing with scientific issues. This is because citizens need to deal with technical, methodological and epistemological scientific uncertainty, which is realised when scientists give differing accounts or interpretations of phenomena. In other words, effective citizens need the skills to evaluate evidence. However, coping with expert disagreement is something that is rarely encountered in school science, according to Christensen. To deal with this, students need to become aware of how scientific knowledge is socially constructed. As well, if literacy were understood in the sense that it currently applies in language literacy, then it would be seen as a social practice that is situated in social situations and social contexts. In this way, emerging citizens would ask the appropriate critical questions such as whose version of science and technology is presented and, as it is applied, whose interests are served. A critical scientific literacy is clearly crucial to effective citizenship, particularly when dealing with complex biotechnological issues.

The way in which biotechnological interventions are framed is Naomi Sunderland's concern. Understanding the framing process provides insight into the relationship between biotechnology, ecology and society. Such a relationship inexorably leads us to see biotechnology as a political and ethical issue. She says that biotechnology as a social practice occurs through the discursive mediating processes of Alienation, Translation, Recontextualisation, and Absorption. To understand this, we must first extend our understanding of media beyond the current conception of television, radio, or print to consider other processes of mediation. Broadly, mediation occurs when meaning moves 'from one text to another, from one discourse to another, from one event to another'. Furthermore, biotechnological mediation must also be understood as a form of political economy. That is, the very foundations of life, something of awe, mystery, spiritual and scientific significance is translated into a product, accessible to few people. Sunderland's prism is the relationship between social practices and discourses. Biotechnology constitutes an array of related practices and discourses. She is concerned that when social practices and discourses become detached from their social environment by discursively constituted boundaries, we are at risk of losing social control over the practices of biotechnology. Sunderland identifies the mediating movements of Alienation, Translation, Recontextualisation, and Absorption that constitute discourse construction.

The concept of alienation is built from the notion of property, which has three characteristics: alienability, rivalry, and excludability. These are present in biotechnological processes when genetic technologies dissociate biological materials from one 'owner' (any living organism) or context to another 'owner' (e.g., "intellectual property") or context (human DNA from the body to a database). Translation, the most overt discursive function, recasts systems of meaning so that an object or process can be recontextualized and recast. It *fixes* meaning in particular ways. At its simplest, biotechnological discourses shift meanings from a "language of life" into technocratic, scientific discourses. Implied in the term, technocratic, is a political and economic aspect, as well as the scientific. Thus, there are biotechnological commercial goods and services, and biotechnological political imperatives. Accompanying this discursive shift is the process of recontextualisation, or the actual material presence of such biotechno-logically-related objects within new social systems and contexts. Such objects include the living organism itself, the site in which it is manipulated, to say nothing of the extensive legal, social, political, financial and economic processes within which it is now enmeshed. Finally, absorption occurs when these biotechnological processes and artefacts enter seamlessly into everyday discourse and practice. In biotechnology, this occurs when the inalienable is commoditised and when the new technology or product becomes the familiar. This is most likely to occur when a process is considered desirable, acceptable, and familiar.

We see in the discourses of government and industry, often rarely distinguishable, a sense of the imperative based on urgency, an unquestioned notion of advance-ment that will secure a better future, and a fear of missing the technological revolution. To question any of this automatically places the questioner in a highly unfavourable subject position, that is, as a reactionary, a zealot, someone who would limit an infertile couple's chance of having a child, or a cancer sufferer from extending their life. Consequently, there is little chance that those who simply ask questions about consequences, intended and unintended, can be heard because the whole apparatus of economy, education, science, commerce, and politics has been 'frameshifted'. Alternate and dissident discourses beat on the outside of very tall and very thick discursive walls.

Bernard McKenna
Business School
University of Queensland

SECTION II CONTEXTUALISING BIOTECHNOLOGY AND NEW MEDIA

PHIL GRAHAM

2. POLITICAL ECONOMIC PROVENANCES
OF BIOTECHNOLOGY

In this chapter, I take political economy to have two distinctly different but interrelated meanings. The first is an approach to understanding human social systems in terms of the ways in which values are produced, exchanged, and distributed (the economic), the ways in which power is produced, distributed and enacted throughout social systems (the political), and how these aspects of human activity interrelate. The second meaning is more strictly etymological: in this view, political economy provides a means of seeing the intersection of *polis* and *oikos*: the complex relations between public and private; between overarching government of a public collective (*polis*) and the private realms of quotidian human existence (*oikos*). These meanings are inextricably linked because the ways in which the mass of people produce values on a daily basis is conducted according to overarching systemic principles.

I take an essentially Marxist approach to the present analysis of biotechnology, treating it as an historical and material phenomenon and investigating it through historical materialist dialectics, an approach to political economic analysis developed by Marx (Fairclough and Graham, 2001). The specific theoretical stance I take is 'cultural realism' which assumes that: the defining values of a culture are expressed through its institutions, media texts, arts, and sciences; that these have real, material bases in human social relationships; that they have real and material effects on individual constituents of a culture; and that those effects are objectively accessible through analyses of relationships between people, their values, their cultural texts, and their institutions (Smythe, 1981: 192-216).

Such an approach would no doubt present difficulties for some Marxist schools of thought since, in the present context, the term "biotechnology" refers almost exclusively to informatic processes. That is to say: contemporary biotechnology consists of firstly abstracting ideal types from decoded DNA and reconfiguring these abstractions to produce changes in the more concrete realm of biological entities—it proceeds from what many Marxists would call "the ideal" to the "real" and is therefore an idealist pursuit that cannot permit of a direct materialist reading (see Jones, 2004).

Yet the very fact of biotechnology's informatic approach indicates that contemporary biotechnology signals a return to the most ancient cultural impulses: at their most advanced, they are not technologies to merely tame or control the external environment; the 'most promising' of biotechnologies is 'gene therapy', an approach that will ideally reorder our biological selves based on informatic

Naomi Sunderland et al. (eds.), Towards Humane Technologies: Biotechnology,
New Media and Ethics, 17–34.

re-engineering (European Union, 1994). Here is what the EU had to say about gene therapy in 1994:

> Scientists generally agree that somatic gene therapy is one of the most promising ways of allowing to alleviate, to cure or to prevent a growing number of genetic as well as acquired diseases, including cancer and even perhaps AIDS. Somatic gene therapy has indeed recently entered the clinical setting as a highly experimental therapeutic procedure. An important and long lasting research effort is still required before routinely performed medical applications can be envisaged. (1994: 1)

Gene therapies remain in the 'most promising' category of biotechnology. That is to say: *ideally* they will provide all sorts of benefits, but nothing much has happened yet in material terms. Such texts are typical of contemporary policy about new technologies – it is typical 'hype' in an age of 'hypercapitalism' (Graham, 2006).

CULTURAL REALISM AND HISTORICAL MATERIALIST DIALECTICS

The approach I take to this work hangs together in the following way: cultural realism is based on the assumption that we can understand cultures through an analysis of social relations, arts, sciences, and institutions, and the texts they produce, and that such texts will give us insight into the values of that culture. Classical dialectics proceeds on the assumption that, on any given topic of any importance, various experts will express differing and sometimes opposing definitions or explanations of the topic (Grote, 1872). The main assumption of historical materialism is that people living together in historically specific social relations shape their reality with the materials they have at hand and, that as these materials change, so to will social relations (Marx and Engels, 1846/1972). By tracing the history of biotechnology as an institutional terminology to describe many different aspects of life, I am attempting not only to show how biotechnology has come to be the set of concepts, practices, and attitudes that it is in today's society, I am also showing the development of the current political economic climate through dialectical investigation of the term "biotechnology"[i] After outlining the genesis of technological trends that long predate the emergence of biotechnology as a distinct term, I show how biotechnology has been used by various institutions throughout the last century.

DEEP-SEATED TECHNOLOGICAL IMPULSES OF "WESTERN" CULTURE

The Judeo-Christian religious tradition pervades and underpins technological developments throughout western societies (Noble, 1997). And political economy is itself an essentially religious and moral project with roots in the mediaeval church (Langholm, 1998). The religious and therefore moral rationales for biotechnology can be traced to the earliest pages of our oldest moral texts.The Book of Genesis contains the following:

Let us make man in our image, in our likeness, and let them rule over the fish of the sea and the birds of the air, over the livestock, over all the earth, and over all the creatures that move along the ground. (NIV Bible, Genesis 1:26).

The development of technologies through which we have achieved external domination has been so dramatic in recent history, Benjamin Franklin's assertion that human beings are "tool-making animals" has become generally accepted as the sole distinguishing characteristic of the human species (Mumford, 1966). The result of this view is a kind of 'technical narcissism' in which all we see in the development of our species is the succession of tools we have used along the way (Mumford, 1966: 108). The opposing view is that the development of a complex symbolic culture is humanity's primary achievement and distinctive feature; that the tools we have made along the way spring from this aspect of our nature; and that to think of the human being 'as primarily a tool-using animal is to overlook the main chapters of human history' (1966: 8). If we take the capacity for mind, culture, and complex symbolism, especially language, as being definitive of the human condition, the primary motivating force in human development, and understand that technologies have merely 'supported and enlarged the capacities for human expression', we can better understand the historical meaning of new technologies (1966: 9). Put briefly, the view that technology leads to complex minds and cultures has entirely different political and moral implications than its opposite: that 'the evolution of language—a culmination of man's [sic] more elementary forms of expressing and transmitting meaning—was incomparably more important to further human development than the chipping of a mountain of hand- axes' (1966: 8). That is because in the "mind first" thesis, aspects of culture are firstly about self-control, self-understanding, and self-expression, whereas the "technology first" thesis suggests that technological development is an unquestionable cultural good because it produces more developed minds. The "mind-first" view of human development suggests a moral imperative for self-control, understanding, and the flowering of expression. The "technology first" thesis contains a moral imperative for the unobstructed development of technology, regardless of human or environmental costs, because it will always be beneficial to the majority of people.

In case my description of the "technology first" thesis seems too strong, let me present some evidence from current research into the early development of humans:

The manufacture and use of early stone tools represents a major evolutionary advance in the behavior of early hominids. Identifying any shifts in brain and cognitive function that may have been associated with this innovative behavior has long been priority in paleoanthropological research. (Stone Age Institute, 2006).

This statement from the Stone Age Institute (2006) makes the assumptions that tool-making modified human intelligence and that behaviour precedes cognitive development. More explicitly, Ambrose (2001) argues that

Human biological and cultural evolution are closely linked to technological innovations. ... Stone tool technology, robust australopithecines, and the genus Homo appeared almost simultaneously 2.5 [million years ago]. Once this adaptive threshold was crossed, technological evolution was accompanied by increased brain size, population size, and geographical range. Aspects of behavior, economy, mental capacities, neurological functions, the origin of grammatical language, and social and symbolic systems have been inferred from the archaeological record of Paleolithic technology. (Ambrose, 2001: 1748)

Ambrose does not posit an explicit causal link between the development of technology and increased capacities for symbolic complexity and cognition, but the assumption that technology—the 'innovative behaviour' referred to by researchers in the Stone Age Institute—leads the process is clear. Understood as behavioural units, the production of complex technologies, argues Ambrose, is technically comaparable to language:

Stone-tipped spears, knives, and scrapers mounted in shafts and handles represent an order-of-magnitude increase in technological complexity that may be analogous to the difference between primate vocalizations and human speech. [...] Assembling techno-units in different configurations produces functionally different tools. This is formally analogous to grammatical language, because hierarchical assemblies of sounds produce meaningful phrases and sentences, and changing word order changes meaning. Speech and composite tool manufacture involve sequences of nonrepetitive fine motor control and both are controlled by adjacent areas of the inferior left frontal lobe. (Ambrose, 2001: 1751)

The basic assumption is made even clearer when Ambrose asserts a relationship between complex tool-making and neurological evolution in the species:

The complex problem solving and planning demanded by composite tool manufacture may have influenced the evolution of the frontal lobe. Functional magnetic resonance imaging demonstrates that the frontopolar prefrontal cortex selectively activates only when imagining a main objective while performing related secondary tasks ... Composite tool manufacture demands the planning and coordination of different kinds of subsidiary tasks and may have coevolved with this frontal lobe parallel processing module. (2001: 1752)

Again, and although co-evolution is proposed, tool-making makes the demands for expanded cognitive capacities. Yet there is a glaring contradiction evident here: Ambrose's conclusion that the development of language and culture might actually be the 'autocatalytic' driver of technological development. It is as if complex 'planning and coordination' could proceed without highly advanced means of cooperation (language), or even that planning could exist without a very complex, communally-shared concept of a future that is susceptible to human planning and coordination. Put differently, Ambrose's whole argument depends on the pre-existence of a socially shared imagination of a future that had not happened yet, and which was strong and clear enough to motivate the production of complex tools. That requires at the very least language with a complex tense system. So much for the "tools first" hypothesis of human culture, communication, and cognition. Yet again and again it appears in "Western" theories of human development: technology leads and humans benefit. With this understanding of "Western" technological traditions in mind, I move on and in the following section specifically to the place of "biotechnology" in recent political economic history.

THE BIOTECH CENTURY: ENGINEERING EVERYTHING FROM BEER TO BABIES TO BODY PARTS

The first recorded mention of biotechnology I have found is by a brewery in Leeds, England (Murphy & Sons, 2007; cf. Bud, 1991, Hulse, 2002). The Bureau of Biotechnology was established in 1899 to investigate the best ways to brew beer by testing various combinations of organisms, metals, and chemicals and to advise other breweries on these matters (Murphy & Sons, 2007; cf. Hulse, 2002). The bureau is later mentioned in Society Proceedings of the *Journal of Parasitology* as having provided a live culture of fungus identified as impeding a 'sheepskin sweating and tanning process' (1925: 218).

It is not surprising that one of the oldest chemical processes in recorded human history (fermentation of alcohol) gets shifted into the realm of science at the turn of the 20th century. It is the period during which the dominant institutions of the West were all refashioning themselves along scientific lines: F.W. Taylor had laid far-reaching claims for "scientific management"; Woodrow Wilson, later elected US president, had coined 'scientific administration' (1887); Dewey sparked generations of debate that continue down to today about what a 'science of education' might mean (Graham & Luke, 2005; Williams, 1959). The new industrial sciences of war had had an overwhelmingly successful applied trial in the American Civil War. Emblematic of this is the Chicago meatworks that was transformed into the first ever production line for Winchester rifles (Standage, 1998).

Given that alcohol was an important part of international trade at the time, and that science was a passion for many in the middle class, the Leeds Brewer who coined the term is very much an historical cipher for the "scientific" spirit of the age. And so it is with subsequent meanings of biotechnology throughout the twentieth century (Bud, 1991). Throughout the course of the twentieth century, argues Bud, and from whichever perspective, biotechnology mediates between biology and engineering (1991): between *what* we do with biology and *how* we do it. And, as David F. Noble puts it, 'the technical work of the engineer' in Capitalism has rarely been anything other than the 'scientific extension of capitalist enterprise' (1977: 33). Noble quotes Henry Towne, who in 1886 says that 'the dollar is the final term in every engineering equation' (1886, cited in Noble 1997: 34). From its earliest formalisation as a widespread and purposive scientific enterprise, "biotechnology" has ultimately been about writing "scripts" for the interaction between machines, bodies, cultures, external environments, and minds. Here is an example from *Science* in 1947:

The capacity, efficiency, and endurance of the human machine in physical labor is the concern of every engineer who assumes administrative duties, and doubly so for the production engineer. These factors underlie the principles of scientific management and time-and-motion analysis. When the additional stresses of environmental temperature, pressure, or anoxia accompany the work, physiological tolerance may be a critical concern. Hours of work, on-the-job feeding, rest periods, etc. are also phases of the physiology of work which form an important part of a comprehensive biotechnology (Taylor & Boelter, 1947: 217)

These enthusiasts of a new biotechnology in 1947 encapsulate the spirit of the preceding five decades. Control of workers had been reaching steadily inwards, past the raw physical aspect of activity, past time and motion studies, into psychology, nutrition, hygeine, and leisure. In this political economic context, biotechnology is directly aligned with strategic management and claims to be its necessary extension. This is a different era of management than the present day. Labour is still understood in classical terms as unruly, *variable* capital (variable capital being the relatively unpredictable amount of work done by labour as opposed to the more predictable "fixed capital", such as machinery and buildings). Today's corporatist managerial class is largely separated from productive factory work which is now typically outsourced and carried out on contract. We can see in the above text that the scientific "pulling apart" of people has begun in earnest. Scientific management was firstly a technique designed to separate thought from action in the labour process; to reduce the variability of variable capital (or labour), separating the process of production into its smallest components; and simultaneously functioning to hide the social character of labour by separating workers from any personal connection with finished products (Smythe, 1981). In this engineering advance, not only is thought separated from action, action and thought are broken down further into their component parts so they can be engineered to conform with the stresses and pressures of physical labour.

The biotechnology program described above, which has been launched at the Engineering school at the University of California, is premised on: 'the interdependence of man and machines', 'the progressive extension of artificial control of human environment', and 'the expanding role of the engineer in human affairs' (1947: 217). It is a thoroughly pedagogical affair, a 'curricular innovation' that is justified on 'practical as well as philosophical grounds', and emphasising the necessity of adding 'a biological phase to the technological equipment of the engineer' (1947: 217). That the representatives of Capital feel it necessary to enclose not merely the physical energies of variable capital, but to take scientific control of psychology, biological processes, and the human environment, indicates that a large-scale move towards the scientific control of attitudes, abilities, and actions had become both feasible and desirable. Again, it is necessary to stress the thread that runs through this article for the journal Science in 1947 and the present hyperbole about the benefits of gene technologies: both have as their aim the abstraction of an ideal type worked out on a scientific basis and written directly into human biology. They differ only in terms of means and extent. It may be superfluous to note that humans, even in 1947, are idealised as mere objects to be manipulated for external ends – as a mere collection of mechanical things to achieve the ends of management and its machines. Three years later we are told that:

> The evolution of a field of human engineering was inevitable. It has as its prime goal the achievement of optimal man-machine relations. It asks the twofold question, How can a man be selected and train to operate the machine most efficiently, and how can the machine be designed so that the man can operate most efficiently? In one instance we select the man; in another, the machine (Mead & Wulfet, 1952: 373).

The situation could not be put any more plainly. By this time, worldwide data were becoming available from anthropology, psychology, and more importantly from a near global system of scientific management coordinated and disseminated through the media of academe (Mead & Wulfet, 1952: 373-378). Mead and Wulfet identify the impetus for an engineering-based, human-focused biotechnology as the greatly increased complexity in the new machinery of war:

> During World War II … it was found that, even with the application of the best selection and classification devices then known, the most up-to-date training techniques, and the most refined techniques of time and motion economy, there were too many instances where military operators were unable to perform the tasks complex modern warfare and its instruments demanded. […] It became clear, for perhaps the first time, that the human being could be the factor that prevented an engineering device from performing to its full specifications … (1952: 373).

Through to the late 1960s, "biotechnology" became shorthand for the science of integrating people into machine systems in the most efficient manner possible according to the lights of engineering and business administration. In essence this conception of biotechnology is neither different to, nor sperate from, the development of scientific management as a set of ideas and practices throughout the 20th century. The prevailing assumptions are: that machines are more reliable than people; that people must be therefore engineered to fit machines; that, as machines become more complex, they are more susceptible to human fallibility; and that increasing complexity of technology is an unquestioned good. The overall implication of this view is total automation.

In the mid 60's ethnicity – or more properly, culture – becomes a barrier to the "proper" engineering of human beings (Pierce, 1966). Again, it is the 'accelerated growth in technological complexity that occurred during World War II' that continues to drive the concept of biotechnology during this period (Pierce, 1966: 218). A strange twist becomes evident at this point: Pierce (1966) argues that it is not the machine component that is constant; rather, 'the design of the human component is fixed, and the type of task it performs is limited by its relatively unmodifiable design' (1966: 218). The overall purpose and orientation of biotechnology at this time remains practically unchanged since the early 1950s:

The fundamental concept of biotechnology is that man should be considered, not as an afterthought to be included only when the major elements of equipment design have been completed, but rather as one of the various components which must be fully integrated into the system. All components, whether human or inanimate, assign certain tasks to perform in order to achieve the over-all goal of the system (Pierce, 1966: 218).

Although it seems that human beings are still considered to be indispensable in any total way, the implications remain the same. A complete statement of the aims of biotechnology at the time is as follows:

To determine the nature and magnitude of the capabilities and limitations of the human component in the man-machine system; to appropriately allocate tasks to the human and inanimate components in accordance with their respective attributes; and to insure [sic] that the interface between man and machine be designed to most efficaciously exploit the distinctive characteristics of each (1966: 219)

Pierce argues that since 'it is now clearly recognized that modern technological equipment can present serious problems in inefficient operation within our own society. It should be obvious, then, in societies which are less technologically sophisticated than ours … such problems could assume proportions which may appear to be virtually insoluble' (1966: 219).

The text by Pierce is interesting for a number of reasons: first, there is a functional inversion of the classical categories of political economy based on *potentials*. Machines are seen as being functionally variable rather than predictable and fixed, whereas people are seen to be functionally fixed in terms of what they can do and how far they can be modified. This tells us much about the tense system at work in respect of technological and human potentials: as far as the future is concerned, the machine's potential is unlimited in the mind of the engineer-administrator. But humans, at least as parts of a machine system, have limited functionality and therefore their involvement in this techno-complex is a hindrance and must be accommodated for at every stage. Even more interesting is the fact that culture begins to play a problematic role for biotechnologists of the day. The implication made by Pierce is that people from technologically less-developed cultures can only increase the problems experienced in the developed world. What is today called "the third world" is a problem to be excluded from biotechnology rather than as the "feed-the-world" moral rationale that it is today.

Aside from various versions of biotechnics – human oriented technologies designed to meet solely human needs, that operate at a human scale, and reintegrate humanity with nature (Mumford, 1934; Bud, 1991) – it is not until 1969 that the environment comes to the fore in definitions of biotechnology (Platt, 1969). By the end of the 1960s a general awareness of looming crises had emerged throughout the developed world, in part as a result of intellectual revolutions during that decade, in part because of the sheer pace at which technology was advancing, and in part because of a global awareness promoted by advanced communication technologies (and superpower posturing) of the sheer destructive power directed at the whole planet on a daily basis (Platt, 1969). Biotechnology, according to Platt, would play a key role in balancing tendencies of the human technological complex with a delicate ecosystem. For Platt, biotechnology would respond to the following problems:

> Humanity must feed into the children who are already in the world, even while we try to level off the further population explosion that makes it so difficult. Some novel proposals, such as food from coal, or genetic copying of champion animals, or still simpler contraceptive methods, could possibly have large-scale effects on human welfare within 10 to 15 years. New chemical, statistical, and management methods for measuring and maintaining the ecological balance could be a very great importance. (Platt, 1969: 167)

Harkins (1975) takes the prevailing winds of a "cultural turn" further, noting that any solutions to the massive problems that had emerged into general consciousness during the 1960s will not be merely technical (1975: 29-31). What the problems require is a complete paradigmatic shift (1975: 27). In this "transitional" text, Harkins sees biotechnology as one of the 'hardware' elements of a sociotechnical complex that needs to be reoriented (1975: 27). The term 'software' is understood in a very specific orientational sense here: 'All software operates to control behaviour. The critical question is: What software for whose behaviour to what ends and with what implications?' (1969: 27). Culture again presents a problem to any clear answer:

> Sociocultural technology in practice becomes a "problem" when paradigms generating past/present/future "realities," desirable and undesirable, do not match up. In our type of society, the critical problem mentioned earlier becomes critical problems as more people become manipulatively, visibly, and widely involved in the cybernetic functions. Software/hardware developments are emerging to "deal with" conflicting stasis and change, but themselves will enter into complex feedback relationships, often of indeterminate dimensions and implications. (1969: 27-28)

Culture, the tense system, technology, an emerging new media complex comprised of computers, and the most ancient tendencies of human social systems must be addressed if the looming crises (extinction, planetary destruction, environmental disasters, unbearable tensions, participatory etc) are to be avoided—nothing less than a global program of re-education is required (1975: 30-32). Platt quite correctly sees education systems as the primary medium through which a global "re-engineering" of values, beliefs, and practices must be achieved (1975: 30-32). He contrasts 'fixed', 'limited learning', and 'open information' systems of education, arguing that the first requirement of the most desirable of these (open information) requires a hardware level that includes a:

> two-way interactive, broad-band communication system connecting every point on the globe with any other end to all space colonies. Each home has its own communication terminal which includes: (1) interactive, multi-channeled

cable television; (2) picture-phone; (3) computer-computer, man-computer, computer-man communication; (4) combination of keyboards, like hands, printers and electronic logic and storage; (5) videotape porta-pack; (6) quad-sound unit including receivers, tape-deck, speakers and turntable. (1975: 32).

In Harkins' ideal system, the whole of humanity would be connected, almost exactly as it is now, with the education system being 'everywhere and nowhere' (32). His exercise in the futurology of cybernetics is worth noting if only because it so accurate in terms of what has come about in the decades since then. Harkins' "hardware" layer for a generalised sociocultural technology suited for a globalised world includes: an 'automated inquiry system' (*eg* Google); an 'automated intelligence system' (NSA's STIC system), a 'talking typewriter' (on which I am currently dictating large sections of this chapter), an 'automatic language translator' (Alta Vista's Babel Fish), 'automatic identification system' that recognises fingerprints or voices (US Homeland Security immigration system), a 'computer psychiatrist' (Eliza), and a 'computer arbiter' that settles disputes (for example, the divorce information kiosks trialled in Australian Family Law Courts) (1975: 33). The total automation of human problematics is ideally solved with the emergence of such a system.

The overall thrust of Harkins' argument is that all these technological advances, as well as generalised access to them, are required for a *new humanism* that can cope with cultural diversity, global education, and an almost complete reorientation of values, including the values of art, music, and literature. In other words, what seems like a progressive, human-oriented turn in the development of technology, culture, and education again requires nothing less than the automation of every aspect of human being. The final item on Harkins' list of technologies required to address the problems created by technologies is a '*Creation and valuation system* ... capable of creative work in such areas as music, art (painting, sculpture, architecture), literature (essays, novels, poetry), and mathematics, and able to evaluate the work of humans' (1969: 32). Although not systematically integrated, the components of such a system exist on a widespread, generally accessible scale today. It is at this point – the point at which a globally organised informatics system becomes both imaginable and technically possible – that preceding conceptions of biotechnology are explicitly conceived of as part of a totalising approach to the control and automation of everything human. The concerns of the cold war era, which include everything from total annihilation to world hunger to unbearable psychological stresses, permit only of technological solutions implemented on a universal basis – the same scale at which the defining conflict of the day operated.

The degree to which cultural and psychological stresses become apparent in intellectual tracts of the day indicates, I would argue, elite concerns with an emergent anarchism, meant here in a very literal sense as an explicit movement to do away with *all* official leadership in favour of an individualistic, self-governing political paradigm. The cultural and political revolutions of the 1960s, which took place under the shadow of two competing superpowers constantly touting their destructive capacities through threats, education, mass-mediated propaganda, political manoeuvering, and raw displays of destructive power in the form of nuclear weapons tests, were almost without exception expressions of a "people power" ethos, epitomised in the behaviour of the 1960s "flower children". The anarchist tendencies of the day were not confined to "the West". In Russia, China, Africa, and the Middle East, political disruptions and societal dissatisfactions were rife. Harkins' informational "revolution" consists of deploying every form of technology to bear upon the control factors, or 'cybernetics', of human society. Machines are presented as not merely more efficient in "material" terms, but as more efficient judges of everything from education to art to politics.

BIOTECHNOLOGY AT THE END OF CAPITALISM

Until the latest of our world conflicts, the United States had no armaments industry. … But now we can no longer risk emergency improvisation of national defense; we have been compelled to create a permanent armaments industry of vast proportions. Added to this, three and a half million men and women are directly engaged in the defense establishment. We annually spend on military security more than the net income of all United States corporations.

This conjunction of an immense military establishment and a large arms industry is new in the American experience. The total influence – economic, political, even spiritual – is felt in every city, every State house, every office of the Federal government. We recognize the imperative need for this development. Yet we must not fail to comprehend its grave implications. Our toil, resources and livelihood are all involved; so is the very structure of our society. (Eisenhower, 1961)

Allan Luke and I have argued elsewhere that the present system can no longer be described as "capitalist" because it has passed into a corporatist phase (Graham & Luke, 2003, 2005; Graham, 2006). The main distinguishing features of capitalism, at least from a Marxist perspective, are its two "great classes": capitalists (owners of the means of production) and workers (Marx, 1976). It has been decades since ownership was the general mode of control for the "ruling classes". What has emerged instead is a system in which the savings of millions have been mobilised to create a global financial system managed by corporate interests, resulting in the separation of ownership from control. Since the mid-1970s when Harkins was writing, the main mechanism for control of people, cities, states, and nations is debt (Graham & Luke, 2005). Eisenhower's farewell speech (1961) was a warning that this emergent system was either on the verge of taking control or had already done so.

Beginning with Watson and Crick's discovery of the DNA double helix in 1953, and following subsequent experiments with RNA conducted with the aid of new computer technologies and optics, the term "biotechnology" became what it is today in around 1980: a term generally taken to mean direct manipulation of genetic structures at the sub-nucleus level of DNA (Galton, 2001). It was precisely 1980 when the biotechnology "gold rush" began:

The date on which molecular biology became big business was 16 January 1980. Reporters had been notified by telegram that are "major announcement" in molecular biology would be made by the company by again and two members of the scientific advisory board, Charles Weissman of the university of Zurich and Walter Gilbert of Harvard. The news delivered at the Boston Park Plaza hotel was that Weissman had cloned and got expression of the human leucocyte interferon gene in biologically active form. (Wade, 1980: 688)

In fact there was absolutely nothing new in the announcement at all, the human leucocyte interferon gene having been successfully manipulated in Japan sometime earlier with the results published in a Japanese journal (Wade, 1980: 688). The real news, as Wade notes, is the biological-corporate nexus that had been formed by the 'mere context of the announcement' (1980: 688). Linking 'the recombinant DNA technique aith the possibility of manufacturing a promising anti-cancer drug' was sufficient 'to produce a major impact on the public imagination', sparking a 'cloning gold rush' that doubled the paper value of the four major biotech companies without a single product being brought to market (1980: 688). The corporate complex had merged with gene technologies and trumpeted their potentials through mass media. The result was a a flurry of financial speculation (Wade, 1980).

It is almost three decades since the Biogen announcement. The corporation has since had five drugs approved by the US Food and Drug Administration (FDA): Rituxan, Zevalin, Tysabri, Avonex, and Fumaderm (Biogen Idec, 2007). It boasts an annual R&D budget of $US750 million with annual revenues of $2.5 billion and has issued just over 340 million shares (valued at $US43.30 at the time of writing for a market capitalisation of around $14 billion) since its initial public offering (Biogen Idec, 2005). Of the corporation's total shares, a mere 1.9% are owned by those who control the corporation on a daily basis, its executives and directors (Biogen, 2006). Biogen has thus survived to become a typical corporatist enterprise: ownership is separate from control and the entire enterprise has become oriented towards public perceptions of its future value, thus becoming primarily a medium of biotech industry "hype" (Graham, 2006).

The 1980 Biogen announcement is the point at which the most basic processes of living matter become exposed to purely commercial interests. This is a fact of importance for any number of reasons some of which I will detail in the following section. It is also emblematic of the direction of corporatism more generally. Its development heralds a further movement "inwards" towards the commodification of more intimate and abstract aspects of life. Between 1947 and 1966, biotechnology was concerned with codifying the intermediate steps between the integration of raw physical labour into an industrial machine system and the integration of intellectual, psychological, cultural, and environmental factors of labour into that system by applications of engineering method. In the age of corporatism, "biotechnology" goes further, concerning itself with patenting new life forms and ancient biological processes in pursuit of future profits; purchasing speculative rights to genetic material on a mass scale; and generally shifting life into the future tense, whether in the form of debt, hopes for longer or never-ending life, or turning the fear of disease, death, and discomfort into a profit motive.

RECENT BIOTECH RUCTIONS IN AN EMERGENT "KNOWLEDGE ECONOMY"

From its ancient outset, engineering has been about harnessing, intensifying, and amplifying natural forces. Almost as soon as "engineering" turns its attention to "machining" basic biological processes for commercial ends, there begins to emerge a flurry of intellectual activity and institutional responses. Predictably, legal studies is among the earliest to enter the fray:

> Recent discoveries in the field of molecular biology that is popularly known as genetic engineering has given rise to a considerable amount of concerning debate both among scientists and in the wider community. The controversy serves to illustrate that genetic engineering experiments are of special importance because they pose unique and unpredictable threats to human life, to the environment and to agriculturally based economies. Ironically, those threats are counterbalanced by important medical, environmental and agricultural benefits ... (Cripps, 1981: 369)

It was about twenty years after Watson and Crick's discovery that the potential for threat from genetic engineering was paid much official attention. Following an outbreak of smallpox in 1973 which began because of an 'accidental release' of the bacteria from a microbiology lab in London, and which caused two deaths before being contained, a flurry of legislative activity began (Cripps, 1981: 369). Against the tide of new and pending regulation, science responded on the grounds of curtailed 'freedom':

> On May 22, 1978, with the promulgation of the Health and Safety (Genetic Manipulation) Regulations, Britain became the first country in the world to regulate genetic engineering by legislative means. The regulations have, however, been severely criticised by members of the scientific community on the basis that they represent *unacceptable inroads into freedom of scientific inquiry*. It has also been claimed that the risks that relate to the use of genetic engineering techniques are not as significant as might have been thought—an argument which is reminiscent of the comments that have been made in the context of the nuclear debate. (Cripps, 1981: 370, my italics).

The advent of commercialised biological engineering is of particular concern for past and future legislation. Self-regulation – the shibboleth of a fast-emerging corporatism – is insufficient, according to Cripps:

> The risks that are involved a particularly acute in the private sector where companies are competing for patents in respect of new products and processes that arrive from the technology of genetic engineering. These companies are manufacturing and using genetically engineered organisms in circumstances which reduce the likelihood of adherence to restraints that lack the force of law. (1981: 371-72).

While legal constraints are necessary, says Cripps, the benefits of genetic engineering cannot be ignored.

It is clear that the benefits that are embodied in genetic engineering lend the technology an urgency which should not be slowed by inefficient administrative controls. It is equally clear that genetic engineering that is conducted in the public and private sectors can be regulated in a manner that is consistent with its development and exploitation and with attempts to safeguard animal and plant populations. (Cripps, 1981: 372)

These two poles of debate – *clear* benefits versus *potential* risks – have basically defined most arguments around the deployment and regulation of genetic technologies ever since. These poles are underpinned by the "progressive" imperative on one side (technological development as unquestionable good) and a "conservative" morality on the other (we should not "play God" with new technologies). Given that the debate is framed in these terms, the outcome is fixed on the side of technology. That is because the term "technology" has come to be identified with the whole of '*modern civilization*' (Smythe, 1981: 217-219):

"Technology" … is said to offer us all kinds of "good" and "bad" things. And when bad things come to pass, more technology in turn will cure them, if we use it to produce more good things, and not more bad things. And so on. … Humanity's ecological crisis will be the result of technology. Developing nations need not high technology but intermediate technology. Catastrophic world war, if it comes, is the result of our use of technology. (Smythe, 1981: 217).

In the future-tense operations of corporatism, the balance between the opportunities and risks of any new technology—assuming they can both be assessed—are invariably rigged in favour of "opportunities" for a number of reasons. First, the financial value of any corporation is now dependent upon market expectations of future performance, which is only to say that a company worth investing in is a company that promises the highest future return on investment. This is a principle that holds the private investor as much as so-called "institutional investors" (Graham, 2006). Second, because of this primary principle, all corporate news is ideally *good* news as a matter of financial sanity. Therefore, any talk of inherent risks in technology must be minimised to ensure that the company maintains its value as perceived by that amorphous entity called "the market". Third, and perhaps most importantly, the term "opportunities" is an integral part of the discourse of unlimited growth, without which the future of any corporation – indeed the entire speculative system – would be in jeopardy (Graham, 2006).

The realm of opportunities "opened up" by biotechnology does not always include overtly pecuniary interests. As soon as genetic engineering became a going concern for corporate interests, it took on a philanthropic mantle:

Developing countries are naturally attracted to the potential applications of biotechnology research in solving problems of hunger, energy supply, and improving the quality of life. The priorities of the different countries vary widely, however. The National Institute of Biotechnology and Applied Microbiology in the Philippines, for example, has accorded priority to research on (i) biofuels; (ii) nitrogen fixation; (iii) food fermentation; (iv) plant hydrocarbons; (v) antibiotics, vaccines, and microbial insecticides; and (vi) biomass production. The National Biotechnology Board of India has chosen genetic engineering, photosynthesis, tissue culture, enzyme engineering, alcohol fermentation, and immunotechnology as areas of interest. Nearly every developing country has plans or programs for harnessing the tools of biotechnology for national development (Swaminathan, 1982: 967)

The rest of this article, again from *Science*, focuses mainly on increased production of food to feed the Third World. This surprising proliferation of Institutes and Bureaux in developing countries so soon after commercialisation exemplifies the propaganda value of "technology" (Smythe, 1981: Ch 10). However, any number of references from the late 1940s onwards could be cited to show that production of sufficient food to feed the whole of humanity had long since been achieved (Horkheimer & Adorno, 1947/1998; Saul, 1997; Lowe, 2007). At the time of writing, two kilograms of food per day per person is produced: more than sufficient to feed everybody (Lowe, 2007). It is not the food production system that needs fixing; it is the food distribution system (Lowe, 2007). Similarly with the development of drugs: the impacts of most serious communicable diseases, whether recent or long-term endemic, have been minimised in the developed world, but in the developing world, the pharmaceuticals required to achieve this remain unavailable due to serious flaws in a distribution system organised solely along monetary lines (Lowe, 2007).

Unfortunately, the philanthropic and humanitarian arguments for the uptake and mass public investment in biotechnology are all too often merely efforts in public relations, or "spin". But this should be surprising in an age of corporatism, where 'public education' about the character and nature of the future—literally, publicly conducted   lessons on how to live better in a permanent future-tense—is an essential part of the system's survival (cf. Price, 1985 on 'the need for public education' about biotechnology). These lessons are taught all the more easily in a system guided by the tenets of technological innovation at practically every level, and evidenced every day by proliferating devices of varying and dubious social worth. As far as "biotechnology" goes, it would appear that the main game for the major corporations involved is control over the *meaning* of the term "biotechnology", a fact they recognised immediately upon the advent of its mass commercialisation (Kleinman & Kloppenberg, 1991). Since commercialisation, the meaning of biotechnology has become subject matter for legal studies (Cripps, 1981), political science (Funke, 1985), language and linguistics (Lassen, 2004), education (Sun, 1981), economics (Siekevitz, 1979), media studies (Hansen, 2006), cultural studies (Harkins, 1975), business (Wade, 1980), philosophy (Thacker, see this volume), ethics (Hulse, 2002), and sociology (Kleinman & Kloppenberg, 1991), (to name but a few), along with the many applied and pure disciplines required to actually implement genetic technologies in any meaningful way.

Children in developing countries can now attend schools that specialise in the teaching of biotechnology.

As part of an alleged "knowledge economy", biotechnology is somewhat of a "superstar": it requires endless pure and applied research that can only be conducted with the aid of massive and complex infrastructure and huge public subsidies; produces patentable material of the most abstract (and therefore elastic) kind; and is most easily and successfully deployed in those areas most fundamental to human survival: food production and health care. It has also sparked endless debate, all of it rigged in favour of the system it represents so fully. Such extensive and heated debate is food for all of us who make a living from intellectual pursuits. As long as the weight of opinion is divided in even a vaguely equal way "for" or "against" "biotechnology", "genetically modified foods", "stem cell research", and so on, debate will continue to flourish, especially as the extremes of the "for" and "against" camps are pushed.

CONCLUDING REMARKS

The corporatist imperative is control of what the future means. The corporate "share" and its future value is quite literally all that matters to those whose work it is to "manage" corporations. Biotechnology, which has come to mean direct control over reproductive processes at the sub-cellular level, can be seen as just one aspect of corporatist control over people and places. From another perspective, the term "biotechnology" has remained little more than a cipher for the political economic systems in which it has been deployed. At the peak of science-as-paradigm (the turn of the 20th century) biotechnology meant nothing more than applying industrial research methods to the age-old process of brewing beer. At the peak of human engineering, between 1911 in 1960, "biotechnology" meant the process of integrating imperfect human beings into the neat, predictable, and unquestionably efficient world of machines. In the so-called post-industrial knowledge economy, biotechnology becomes a medium for systemic propaganda; it has become part of the daily curriculum to which the public is exposed in order to fit itself into an idealised future produced by a system that is entirely future-oriented and reliant upon new technologies.

The history of the term "biotechnology", short though it is, reaches back into the depths of political economy, human history, and human culture. It can find legitimacy in the opening chapters of the Old Testament; precursors in the most ancient human techniques of selective breeding and fermentation; philosophical and political rationales in Plato's idealism (Galton, 2001); and scientific rationales from late scholasticism to the present day. Most of all, and at the very deepest level, it reminds us that our first moves as self-conscious, culturally active beings was to write upon our own bodies, adding to and removing from our bodies: decorating, deforming, beautifying, and mutilating ourselves. The term is not merely a cipher; it is a reminder of our cultural and artistic origins and their relationships with what we see as beautiful, normal, desirable, and powerful. It reminds us that we are material beings with propensities for idealism, and that we make our environments, as far as possible, in our own likeness.

NOTES

ⁱ A similar approach has been called 'discourse historical' analysis (Reisgl and Wodak, 2001). It differs because it does not focus specifically on culture, technology, and political economy

REFERENCES

Ambrose, S. H. (2001). Paleolithic technology and human evolution. *Science, 291*, 1748–1753.
Biogen Idec. (2007). *About Biogen Idec*. Retrieved April, 2007, from http://www.biogenidec.com/site/013.html
Bud, R. (1981). Biotechnology in the twentieth century. *Social Studies of Science, 21*(3), 415–457.
Cripps, Y. (1981). A legal perspective on the control of the technology of genetic engineering. *The Modern Law Review, 44*(4), 369–387.
Eisenhower, D. D. (1961, January 17). *Farewell radio and television address to the American people by president Dwight D. Eisenhower*. Kansas: The Dwight D. Eisenhower Library. Retrieved December, 2004, from http://www.eisenhower.archives.gov/farewell.htm
Fairclough, N., & Graham, P. (2002). Marx as a critical discourse analyst: The genesis of a critical method and its relevance to the critique of global capital. *Estudios de Sociolinguistica, 3*(1), 185–229.
Funke, O. (1985). Biopolitics and public policy: Controlling technology. *BioScience, 32*(6), 486–487.
Graham, P., & Luke, A. (2005). Militarising the body politic: Feudal influences in the language of Bush, Blair, and Howard. *Language & Politics, 4*(1), 11–39.
Grote, G. (1872). *Aristotle* (Vol. 1). London: John Murray.
European Union. (1994). Opinion on the group of advisers on the ethical implications of biotechnology to the European Commission. *4*. Retrieved December, 2006, from http://ec.europa.eu/european_group_ethics/docs/opinion9_en.pdf
Galton, D. (2001). *In our own image: Eugenics and the genetic modification of people*. St Ives: Little, Brown and Company.
Graham, P. (2006). *Hypercapitalism: Language, new media, and social perceptions of value*. New York: Peter Lang.
Hansen, A. (2006). Tampering with nature: 'Nature' and the 'natural' in media coverage of genetics and biotechnology. *Media Culture Society, 28*(6), 811–834.
Harkins, A. M. (1975). Controls, paradigms and designs: Critical elements in the understanding of cultural dynamics [special section on structures of censorship, usually inadvertent: Studies in a cultural theory of education]. *Council on Anthropology and Education Quarterly, 6*(2), 27–34.
Horkheimer, M., & Adorno, T. (1947/1998). *Dialectic of enlightenment*. London: Verso.
Hulse, J. H. (2002). Ethical issues in biotechnologies and international trade. *Journal of Chemical Technology and Biotechnology, 77*, 607–715.
Jones, P. E. (2004). Discourse and the materialist conception of history: Critical comments on critical discourse analysis. *Historical Materialism, 12*(1), 97–125.
Kleinman, D. L., & Kloppenberg, J. (1991). Aiming for the discursive high ground: Monsanto and the biotechnology controversy. *Sociological Forum, 6*(3), 427–447.
Langholm, O. (1998). *The legacy of scholasticism in economic thought: Antecedents of choice and power*. Cambridge, MA: Cambridge University Press.
Lassen, I. (2004). Ideological resources in biotechnology press releases: Patterns of theme/rheme and given/new. In L. Young & C. Harrison (Eds.), *Systemic functional linguistics and critical discourse analysis: Studies in social change* (pp. 264–279). London: Continuum.
Lowe, I. (Lecturer). (2007, February 11). *Big ideas: Rick Farley lecture series* [radio broadcast]. Sydney: ABC Radio National. Retrieved March, 2007, from http://www.abc.net.au/rn/bigideas/stories/2007/1842533.htm
Marx, K. (1976). *Capital: A critique of political economy* (Vol. 1, B. Fowkes, Trans.). London: Penguin.
Marx, K., & Engels, F. (1846/1972). The German ideology. In R. C. Tucker (Ed.), *The Marx/Engels reader*. New York: W.W. Norton.

Mead, L. C., & Wulfet, J. W. (1952). Human engineering: The study of the human factor in machine design. *The Scientific Monthly*, *75*(6), 372–379.

Mumford, L. (1934/1962). *Technics and civilization*. New York: Harcourt Brace.

Mumford, L. (1966). *Technics and human development: The myth of the machine volume one*. New York: Harcourt Brace & Javanovich.

Murphy & Sons. (2007). *About us*. Leeds: Murphy & Sons. Retrieved December, 2006, from http://www.murphyandson.co.uk/aboutUs.htm

NIV Bible. (2006). *Genesis*, *1* p. 26. Retrieved December 2006, from http://www.ibs.org/niv/passagesearch.php?passage_request=Genesis%201&niv=yes

Noble, D. F. (1997). *The religion of technology: The divinity of man and the spirit of Invention*. New York: Alfred A. Knopf.

Pierce, B. (1966). The ethnic factor in biology. *Economic Development and Cultural Change*, *4*(2), 217–229.

Platt, J. (1969). What we must do. *Science*, *166*(3909), 1115–1121.

Price, H. S. (1985). Biotechnology: The need for public education. *Bioscience*, *35*(4), 211.

Reisgl, M., & Wodak, R. (2001). *Discourse and discrimination: Rhetorics of racism and antisemitism*. London: Routledge.

Siekevitz, P. (1979). Biotechnology and profit. *Science*, *206*(4424), 1257–1258.

Smythe, D. (1981). *Dependency road: Communications, capitalism, consciousness, and Canada*. Norwood, NJ: Ablex.

Standage, T. (1998). *The Victorian Internet*. New York: Berkley Publishing Group.

Stone Age Institute. (2006). *Research*. Retrieved December, 2006, from http://www.stoneageinstitute.org/c_research.shtml

Sun, M. (1981). New patent rule upsets universities. *Science*, *213*(4513), 1234–1235.

Swaminathan, M. S. (1982). Biotechnology and third world agriculture. *Science*, *218*(4576), 967–972.

Taylor, C. L., & Boelter, L. K. M. (1947). Biotechnology: A new fundamental in the training of engineers. *Science*, *105*(2722), 217–219.

Wade, N. (1980). Cloning gold rush turns basic biology into big business. *Science*, *208*(4445), 688–692.

Williams, J. (1959). Dewey and the idea of a science of education. *The School Review*, *67*(2), 186–194.

Wilson, W. (1887). The study of administration. *Political Science Quarterly*, *2*(2), 197–222.

Phil Graham
Creative Industries Faculty
Queensland University of Technology

EUGENE THACKER

3. DATA MADE FLESH

Biomedia and the Body

...and all around you the dance of biz, information interacting, data made flesh in the mazes of the black market. (William Gibson, *Neuromancer*)

NEW BODIES, NEW MEDIA?

The aim of this essay is to elaborate some of the consequences of the intersection between information technologies and biotechnologies. I will be using the term "biomedia" to designate this intersection, one in which a certain isomorphism is produced between genetic and computer "codes." As we will see, biomedia do not simply involve the "computerization" of biology, in which the biological domain is rendered as informational and immaterial. In fact, central to biomedia is a notion of "life" that is at once informatic and yet not immaterial. Before considering biomedia as a concept in itself, however, it will be necessary to understand how the relation between the body and technology has been understood in a postmodern, post-industrial context. We will therefore consider the various ideologies concerning a technologically-mediated "posthumanism" before considering the relation between this posthumanism and the biotech industry. It will be in the space between these that the concept of biomedia will become important.

Over the past few years, it has become increasingly commonplace to come across a new vocabulary in mainstream media reportage: headlines about "genomes," "proteomes," "stem cells," "SNPs," "microarrays," and other mysterious biological entities have populated the many reports on biotechnology. The completion of human genome projects, policy decisions concerning the use of embryonic stem cells, controversies over genetic patenting, and the ongoing debates over human therapeutic cloning, are just some of the issues which biotech research brings to public discussion. For many advocates as well as detractors, the so-called "biotech century" appears to be well underway.

But we might be a little more specific in characterizing biotech, which is for many becoming the new paradigm in the life sciences and medical research. This is to suggest that one of the main things which characterizes biotech currently is an intersection of bio-science and computer science, or to put it another way, an intersection between genetic and computer "codes." Within biotech research, this is known as the field of "bioinformatics," which is simply the application of computer technology to life science research. Its products include online genome databases, automated gene sequencing computers, DNA diagnostic tools, and advanced data

Naomi Sunderland et al. (eds.), Towards Humane Technologies: Biotechnology, New Media and Ethics, 35–48.

mining and gene discovery software applications. When we consider advances in these fields, it becomes apparent that what characterizes biotech is a unique relationship between the biological and the informatic. As Craig Venter, CEO of Celera Genomics states, "We are as much an infotech as a biotech company"; a notion reiterated by Ben Rosen, chairman of Compaq Computing, who states that "biology is becoming an information science."

The questions which these mergers between biotech and infotech bring up are many: What does it mean to have a body, to be a body, in relation to genome databases? How is our notion of the body transformed when biotech research demonstrates the ability to grow cells, tissues, and even organs in the lab? How is the boundary between biology and technology reconfigured with the DNA chips commonly used in biotech labs? In biotech research, what happens to the referent of "the human" as it is increasingly networked through information technologies?

Between biology and technology, genetics and computer science, DNA and binary code, is a more fundamental relationship between human and machine. In this paper I will take "posthumanism" as a wide-ranging set of discourses which, philosophically speaking, contain two main threads in its approach to the relationship between human and machine. The first thread I will refer to as "extropianism," which includes theoretical-technical inquiries into the next phase of the human condition, through advances in science and technology. These are mostly technophilic accounts of the radical changes which leading-edge technologies will bring. The second thread is a more critical posthumanism, often in response to the first, and includes key texts by contemporary cultural theorists bringing together the implications of postmodern theories of the subject and the politics of new technologies.

Both threads offer valuable insights into the ways in which notions of "the human" diversify, self-transform, and mutate as rapidly as do new technologies. The first part of this paper will be spent analyzing and critiquing the extropian branch of posthumanist thought, especially as it relates to the ways in which the term "information" is defined. The second part will consider research in bio-technology in light of posthuman discourses. While biotech research raises many of the issues common to both the extropian and critical posthumanist discourses, it also elucidates unique relationships between human and machine, flesh and data, genetic and computer "codes."

EXTROPIAN INVASIONS

There is a growing body of research, both theoretical and practical, on the ways in which advanced technologies - from nanotechnology to neural computing - will enhance, augment, and advance the human into a posthuman future. Scientist-theorists, such as Hans Moravec, Ray Kurzweil, Marvin Minsky, and Richard Dawkins, have all been associated with this line of thinking. Organizations such as the Extropy Institute and the World Transhumanist Organization have also been instrumental in creating networked communities based on transhumanist and extropian ideas.

One salient feature of such transformations includes the concept of "uploading," in which the parallels between neural pattern activity in the human mind, and the capacity of advanced neural networking computing, will enable humans to transfer their minds into more durable (read: immortal) hardware systems (Moravec, 1988, pp.109-110). All of this is made possible via a view of the body which places special emphasis on informational patterns. Once the brain can be analyzed as a set of informational channels, then it follows that those patterns can be replicated in hardware and software systems. As Ray Kurzweil states:

> Up until now, our mortality was tied to the longevity of our hardware. When the hardware crashed, that was it. For many of our forebears, the hardware gradually deteriorated before it disintegrated. As we cross the divide to instantiate ourselves into our computational technology, our identity will be based on our evolving mind file. We will be software, not hardware...the essence of our identity will switch to the permanence of our software (1999, pp. 128-129).

Other changes include the transformation of the material world, including the biological domain, through nanotechnology (the construction of organic and non-organic objects atom by atom, molecule by molecule), new relationships to the environment through biotechnology, and the emergence of intelligent computing systems to enhance the human mind.

A key feature of this type of posthumanism - what I'll be generally referring to as "extropianism" - is that it consciously models itself as a type of humanism. That is, like the types of humanisms associated with the Enlightenment, the humanism of extropianism places at its center certain unique qualities of the human - self-awareness, consciousness and reflection, self-direction and development, the capacity for scientific and technological progress, and the valuation of rational thought. As Max More's "Transhumanist Declaration" cites, key principles include "perpetual progress," "self-transformation," "practical optimism," "intelligent technology," "open society," "self-direction," and "rational thinking." (More, retrieved from the Extropy Web site, http://www.extropy.org).

Like the Enlightenment's view of science and technology, extropians also take technological development as inevitable progress for the human. The technologies of robotics, nanotech, cryonics, and neural nets all offer modes of enhancing, augmenting, and improving the human condition – these are also pre-modernist; especially when they refer to a technologically produced "afterlife" or "life ever after" (see also Isaacs' discussion of premodernity and the good in this volume). A key element in the extropian approach towards technology is that technological progress will necessarily mean a progress in "the human" as a species and as a society; that is, just as the human will be transformed through these technologies, it will also maintain, assumedly, something essential of itself. It is in this tension between identity and radical change, between visions of software minds and the realities of biological bodies, that extropianism reveals the inner tensions of posthumanist thinking – it is also a rather typical reaction to new technologies: Cartesian conceptions of body as pneumatic mechanism; Newtonian views of the mind-as-clockwork, etc etc. As a particular thread in the discourse of the posthuman, extropianism can be characterized along three main lines: as a technologically-biased revision of European humanism, as an approach to technology as both self and not-self, and as a tendency to apply life science concepts towards social and political problematics.

For example, the blindspot of this thread of posthumanism is that the ways in which technologies are themselves actively involved in shaping the world are not considered. To borrow a term from Bruno Latour, extropianism privileges the technologically-enabled subject as the agent of change without due consideration to the ways in which "nonhumans" and "actants" are also actively involved in the transformation of the world (Latour, 1999, pp. 122-123). This is not to suggest that we somehow invest our technologies with human subjectivity, but that the situated, contingent effects of technologies are indissociable from the subjects that "use" those technologies, and from the social, political, and economic environments they inhabit. While the humanist slant of extropian thinking clearly privileges a futuristic vision serving the human (post-biological life in hardware systems, intelligence-augmented minds, and more close to the present, extended life span, genetically-modified health, and smart drugs), what remains unclear for the extropians is the extent to which the human can be transformed and still remain "human."

Extropianism escapes this problem by claiming a universality to certain attributes, such as reason, intelligence, self-realization, egalitarianism, ethical thinking, and transcendence – it is an essentially idealist conception of the human, divorced from materiality, history, and context. By assuming that "intelligence" and "sentience" will remain constants over time, and through successive transformations, extropianism smuggles humanist-based conceit into a technologically-driven evolutionary paradigm. The conflicts arise when posthuman thinkers must consider the fate of the human, or its history. What often goes unconsidered are the ways in which the human has always been posthuman, and the ways in which technology has always operated as a nonhuman actant. This looks to be an interesting point. (Ansell Pearson, 1997, pp. 123-150; Latour, 1999, pp. 174-216). In addition, one of the crucial requirements for the posthuman is that technology be approached first and foremost as a tool. This technology-as-tool motif - an investment in enabling technology - operates in several ways. In one sense it presupposes and requires a boundary management between human and machine, biology and technology, nature and culture. In this way extropianism necessitates an ontological separation between human and machine. It needs this separation in order to guarantee the agency of human subjects in determining their own future, and in using new technologies to attain that future. This is an asymmetrical separation, in which the human subject is the actor, and the technology the prosthetic which the human subject uses (Hayles, 1999, pp. 2-3).

This separation also provides the assurance of the neutrality of technology. As Marshall McLuhan long ago (and Mumford long before him) argued, the most dangerous position vis-a-vis technology is to assume its neutrality (1964/1995, pp. 11-12). In this way, the safe, secure space of pure research can provide for a range of utopian possibilities without regard to the historical, social, and political contingencies which enframe each technological development. Thus the human - or rather a humanist standpoint - becomes the safeguard against the threat of technological determinism. It is the human user that guarantees the right, beneficial use of otherwise value-neutral technologies.

Thus technology operates in a complex way for the extropian branch of post-humanism. It is taken as a tool, and one that is both transparent and value-neutral. It is thus abstracted from any sociohistorical contingencies. But this ontological separation also hides a fantasy of technology embedded in the posthuman generally: that fantasy is one about the anachronism of technology, in which the human advances so far that it doesn't need technology, that technology in effect disappears. The goal here is to attain a state of optimum self-sufficiency, autonomy, and self-realization such that the management of the human/machine divide is, in fact, no longer necessary. While in one sense this would seem tantamount to saying that the human becomes technology, the rhetoric of extropianism is - like that of most technophilic movements - about the world in the service of the human (be it the natural world, as in biotech, or the artificial world, as in AI).

OTHER POSTHUMANISMS

One of the most resonant aspects of Donna Haraway's 1985 "Cyborg Manifesto" was that its appropriation of the terminology of the cyborg was itself a perform-ative gesture against the necessity of origin stories (pp. 180-181). By strategically borrowing the figure of the cyborg from Space Race-era NASA research into enabling astronauts to survive in "alien" or extraterrestrial environments, Haraway shows how the doubled contingency of humans and technologies will always require critical gestures, ironic gestures, even ludic gestures, which will turn upside-down, and render impure and non-innocent, our views of the human condition.

This move is also a key element to the more critical threads of posthumanist thinking, which are often interventions in the overtly utopic postulations of thinkers like Moravec and those associated with extropianism above. Thinkers such as Haraway, Katherine Hayles, Rosi Braidotti, and Cary Wolfe have shown how any critical perspective on the human-technology relationship will have to pay special attention to the underlying assumptions in place in declarations such as those by the Extropy Institute. While not denying the significance and transformative possibilities of new technologies, these critical takes on the posthuman offer a more rigorous, politically and socially rooted body of work, from which the difficult task to imagining the future may begin.

While Haraway's focus has primarily been the life sciences, Katherine Hayles (1999) has offered several pointed, detailed analyses of posthumanist thinking (in its extropian vein). Focusing on research in advanced computing and cybernetics (AI, robotics, emergence, cognition), Hayles shows that the posthuman is founded on a strategic definition of "information." This modern notion of information - most notably in the extropian concept of "uploading" - does not exclude the body or the biological/material domain from mind or consciousness, but rather takes the material world as information. This powerful ideology not only informs research in cognitive science but in the life sciences as well. Hayles' critical point is that informatics is a selective process, and those things which are filtered or transformed in that process - such as a notion of the phenomenological, experiential body, or "embodiment" - simply become by-products of an informatic economy.

Both Haraway and Hayles have taken up the discourse of the posthuman, and have provided articulate analyses and critique, while not totally denouncing posthumanism itself. The result, as with Haraway's strategic appropriation of the cyborg, is a new hybrid discourse which emphasizes the productive tensions between contingency and emergence. For Haraway, the posthuman can become a unique type of politics, challenging the ways in which the relationships between humans and nonhumans, and biology and technology, are all regulated. As Ira Livingston and Judith Halberstam state:

> The posthuman does not necessitate the obsolescence of the human; it does not represent an evolution or devolution of the human. Rather it participates in re-distributions of difference and identity. The human functions to domesticate and hierarchize difference within the human (whether according to race, class, gender) and to absolutize difference between the human and nonhuman. The posthuman does not reduce difference-from-others to difference-from-self, but rather emerges in the pattern of resonance and interference between the two (1995, p. 10).

It is this processual character of the posthuman that Haraway, Hayles, and Livingston and Halberstam highlight, a zone of transitionality, which does not take its legitimation from any origin, and which interrogates the technological determinism implicit in extropian-type thinking. But for all this, the transitional, transformative, mutating potential of the posthuman is not simply a free-floating, abstract "rhizome." As Haraway makes clear, the posthuman can only work as a biopolitics if it constantly questions what comes to us as "second nature." Part of this work means interrogating and creating the possibilities for the emergence of new relationships between human and machine, biology and technology, genetic and computer information. This in turn means that a consideration of the historical trajectory of information technologies must form an essential part of any understanding of biotechnology.

In her work on the technical genealogies of cybernetics and posthumanism, Hayles locates the emergence of a technologically-derived episteme associated with the information theory of Claude Shannon, and the cybernetics of Norbert Wiener. In *The Mathematical Theory of Communication* (1949), Shannon and Weaver provide the technical foundations for modern communications technologies by conceiving of a unilinear transmission line (a message transmitted from A to B). Likewise, in his equally technical treatise *Cybernetics* (published in 1948), Norbert Wiener established a mode of thinking of machines or organisms as relay-systems which incorporate feedback, input, output, and noise (Wiener, 1948/1996). It is in this tradition that Hayles proposes a shift, from more traditional, modern notions of subjectivity based on presence and absence (we are reminded here of Descartes' criteria of a mind present to itself), to an episteme based on a related dichotomy between "pattern" and "randomness" (Hayles, 1999, pp. 39-40).

At issue in each dyad (presence-absence; pattern-randomness) is a hierarchical valuation, but central to the shift itself is an increasing acceptance of a world-view based on an essentializing of information as the source of an object. For Hayles, the danger with the shift to pattern and randomness is that it contains the potential to simply replay the ideologies and anxieties of the presence/absence dyad, resulting in a devaluation of the body and materiality, and a valuation of the manipulability, replicability, and disembodiedness of information.

The assumption which Shannon and Weiner work from is that meaning is and should be stable with regards to information. However in order to secure such stability the transmission of meaning must also be stable, meaning that the carriers of information, the transmission of information, must also be stable, constant, and thus transparent. This is not a theoretical question, but a technical question, a question of operationality and systematicity (and it was about voice quality over phone lines in particular). Ironically then, in order to secure the stability of information as meaning, researchers in computer science, information theory, and cybernetics must also focus on the transmission, carriers, and the encoding/decoding processes of information. The question for Shannon and Wiener is "how can we keep such-and-such a medium from affecting the meaning of the information signal?", and not "how will such-and-such a medium affect the meaning of the information signal?"

The very language of computer science contains within it this assumption; signals may be encoded, transmitted, and decoded across a range of media, as long as the media are technically able to facilitate the transmission of information that is self-identical. Thus the questions which Shannon and Wiener separately ask result in their theoretical formulations: for Shannon and Weaver working on telecommunications problems at Bell Labs, information is a quantitative measure of the accuracy of the reproduction of a signal from point A to point B (Shannon & Wiener, 1965, pp. 8-16). For Norbert Wiener, working at MIT and for the military, information is the range of "choices" available at a particular instant, within a cybernetic system composed of inputs/sensors, outputs/effectors, and a central mechanism of "feedback."(1948/1996, pp. 6-9). Both researchers ground their research in a notion of information as (i) concurrent with meaning but stabilized through a medium, (ii) a quantitative value independent of qualitative changes or changes in meaning, and (iii) a value stable across media and therefore independent of media. These characteristics, which form what we might call a "classical theory of information," are directly related to the ways in which the posthuman has traditionally equated information with disembodiedness (Hayles, 1999, pp. 4-5, pp. 47-48).

While these are not problematic implications in themselves, when taken within the larger context of the relationship between information technologies and technoscience, they replay the association between disembodiedness and information characterized by Hayles. The reason information can be a self-identical value, across media, across signifying processes, and across systemic contexts, is precisely because it is conceived, from the beginning, as a value independent of material instantiation or of culture. When information is regarded as information, no matter what medium "carries" it, it then becomes a universal, disconnected from the material-technical necessities of the medium, the processes, language, culture, and context. It is this universalizing and de-contextualizing of information that enables Wiener to conceive of machines and organisms as the same, from the perspective of cybernetic systems operating through feedback loops. I do not want to imply here a critique of Wiener's overall suggestions regarding cybernetic systems; rather, it is the particular way in which information - the central unit of Wiener's and Shannon's theories it is the way in which it is or is not intimately constrained by the contingencies of embodiedness that provides the point of problematisation: the theory of information which these foundational texts present to us is one in which information is universalized, de-contextualized, and disconnected from the necessities of technological contingency.

It is this process of making a certain definition of "information" foundational to considerations of the body - which I am locating in the work of Wiener and Shannon - which we might refer to as "informatic essentialism." Informatic essentialism is not a repression, denial, or effacement of the body - rather, it proposes that the relationships between the biological body and information technology is such that the body may be approached through the lens of information. In other words, by making informatics a foundational world-view, the body can be considered as "essentially" information; the product of informatic processes as well as susceptible to being perfectly rendered as information. The complexity in the posthuman position outlined here is that, on the one hand, it does not necessarily deny materiality or the body, but on the other hand, in equating information with the body it only interprets materiality and body in terms of informational pattern. Change the code, and you change the body.

FROM BIOTECH TO BIOMEDIA

What we have not accounted for is the ways in which current developments in the life sciences are equally active in the material transformations of notions of the body and "life itself." This inquiry, this investigation into the informatic qualities of the biological body, is already taking place in contemporary molecular biotechnology, through the immanently practical means of research, clinical trials, product pipelines, and medical application. In press releases from biotech corporations, in articles in science publications, in interviews with researchers, one increasingly hears a refrain. As Nobel-laureate and genomics pioneer Leroy Hood puts it, "biology is information" (retrieved from the Biospace Web site, http://www.biospace.com). Emerging fields, from proteomics to regenerative medicine, are employing computer technology and computer science research into the "wet lab."[i] Such practical transformations assumedly bolster the biotech industry by making genome mapping, gene targeting, and product development more efficient. But on the research side, such intersections between bio-science and computer science may also significantly transform some of the foundational concepts in molecular genetics. For instance, the initial report of the human genome map revealed, among other things, that the number of human genes was far less than researchers had expected, thus prompting many within the research community to call for more "complex" approaches to studying gene expression, biopathways, and biological "systems" (see also Bonfiglioli this volume). Similarly, the controversies over a number of population-genome projects (most notably in Iceland) have raised issues over how ethnicity and race are assumed to smoothly overlap with culture - all of which is being interpreted through genetic data.

In contradistinction to the discourses of posthumanism which seek to dematerialize the body (into software Minds, into informational networks), research in biotechnology presents us with a case in which informatic essentialism is utilized to redefine biological materiality. Biotech assumes the classical definition of "information" and informatic essentialism, but instead of using this definition to direct itself towards the immanence of "disembodied pattern" (to borrow Hayles' terms), biotech begins to reconfigure the materiality of the body through the lens of technology. In doing so, it is formulating and renegotiating new norms concerning how bodies will be approached by the life sciences and medical practice. That norm takes different forms in different contexts, but in general it has to do with (i) a body that can be effectively approached on the level of information, (ii) a body that, as information, can be technically manipulated, controlled, and monitored through information technologies, and (iii) most importantly, a body that is viewed as fundamentally "information" (genetic "codes"), where its being viewed as information does not exclude its being material. This last point is crucial, because it points to the disconcerting ways in which biotech demands that bodies be both informatic and material.

To put it another way, biotech has no body-anxiety. In fact, it is based on a deep investiture and revaluation of the body as a materiality, and one that can be understood and controlled through "information." Biomedical science enframes this as the recuperated, healthy, homeostatic body - a return to its state of health. But the process is less a circle than a kind of spiral: the body returning to itself is fundamentally different from itself, because it has been significantly re-mediated through genetics, gene therapy, stem cell engineering, and so forth. The upward part of this spiral is a self-sufficient, autonomous, immortal body - the dream of the liberal-humanist subject as black-box. The downward part of the spiral is the expendable, unstable body - the fears of the loss of autonomy associated with differentiation, otherness, and expendability. Biotech is, above all, a discourse of production and materialization with respect to the scientific body.

As a way of analyzing this further, we can take one particular field as a kind of case-study, a field within biotech research known as "regenerative medicine" (Bonassar & Vacanti, 1998; Mooney & Mikos, 1999; Petit-Zeman, 2001). Primarily a response to the overwhelming demand for tissue and organs in transplantation, regenerative medicine encompasses research in tissue engineering and stem cells, as well as borrowing techniques from therapeutic cloning, gene therapy, and advanced surgical techniques. Its goal is to be able to regenerate and synthesize biological tissues, even entire organs, in the lab. This new horizon of what researchers call "off-the-shelf organs" has prompted many in the medical community to envision a future in which the body's natural capacity to heal itself is radically enhanced through molecular genetics and cellular engineering – an outering of the immune system. Already several products, including a bio-engineered skin graft, are being marketed by biotech companies under FDA approval, and laboratory animal experiments involving the synthesis of a tissue engineered kidney, liver, and even heart are currently underway. Recently, regenerative medicine has made headlines for its discovery of "adult stem cells," or cells within the adult body that contain the potential to differentiate into a wide range of cell types, pointing the way for further research into diseases such as Parkinson's and Alzheimer's (Hall, 2000).

On the one hand, the notion of growing organs in the lab evokes the kind of medical horror often seen in science fiction, from Mary Shelley's *Frankenstein*, to the early films of David Cronenberg. On the other hand, regenerative medicine is promising to be among the first medical fields to be able to turn the knowledge (and data) generated by biotech into practical medical application. Using the techniques of regenerative medicine as our example, we can see three primary moments which characterize this intersection between biotech and infotech:

The first has to do with the "translatability" between flesh and data, or between instantiated genetic codes and computer codes. In order for a patient to receive a bioengineered skin graft, blood vessel, or cartilaginous structure, a biopsy or cell sample must first be taken. Using genetics diagnostics tools such as DNA chips and analysis software, DNA samples are translated into computer codes that can be analyzed using bioinformatics software. That is, once the biological body can be effectively interpreted through the lens of informatics, then a unique type of encoding can occur between genetic and computer codes. This first step of "encoding" the biological into the informatic is one of the defining moments in the posthuman, allowing the necessity of material instantiation to give way to the mutability of computer code.

The second manner in which biotech integrates itself with infotech is through a technique of programming or "recoding." One of the main breakthroughs which has enabled tissue engineering to regenerate tissues and organs, has been the research done into stem cells. Briefly, stem cells are those cells which exist in a state of pluripotency, prior to cellular differentiation, in which they may become, for instance, bone, muscle, or blood cells. Researchers can target specific gene clusters that might be activated or de-activated for regeneration to occur. All of this takes place through software applications and database tools that focus on the multiple genetic triggers which take a stem cell down one route of differentiation or another.[ii] Once the biological body can be effectively "encoded" through informatics, then it follows that the re-programming of that code will effect analogous changes in the biological domain.

Finally, regenerative medicine mobilizes these techniques of encoding and recoding towards its output - or "decoding" - which is the use of an informatics-based approach to generate or synthesize biological materiality. This is the main goal of tissue engineering: to be able to use the techniques of biotech to actually generate the biological body on-demand – just-in-time spare parts. Once a patient's cells can be prompted to regenerate into particularized tissue structures, they can then be transplanted back onto the body of the patient, in a strange kind of biological "othering" of the self. From the perspective of medical research, this process is purely "natural," in the sense that it involves the integration of no non-organic components, and in the sense that it utilizes biological processes - in this case, cellular differentiation - towards novel medical ends.

In the research of regenerative medicine, this three part process of encoding, recoding, and decoding the body operates through a kind of informatic protocol in which, at each step, information comes to account for the body (see also Sunderland's discussion of 'recontextualisation' in this volume). It is this process that I refer to as "biomedia." Put briefly, biomedia establishes an equivalency between genetic and computer codes – such that the biological body gains a novel technics. The significance of this technical mobility has been described by Donna Haraway:

> ...the genome is an information structure that can exist in various physical media. The medium might be the DNA sequences organized into natural chromosomes in the whole organism. Or the medium might be various built physical structures, such as yeast artificial chromosomes or bacterial plasmids, designed to hold and transfer cloned genes...The medium of the database might also be the computer programs that manage the structure, error checking, storage, retrieval, and distribution of genetic information for the various international genome projects that are under way. (1997, p. 246).

While both research into Artificial Intelligence and biotechnology participate in the assumptions regarding an informatic basis to the body, the primary difference is that biotech directs its resources towards an investment in generating materiality, in actually producing new bodies through informatics. If areas such as genomics and bioinformatics are predominantly concerned with programming the (genetic) body, other areas such as tissue engineering and stem cell research are predominantly concerned with being able to grow cells, tissues, and even organs, *in vitro*, *in silico,* and *in vivo*. The trajectory of biotech's informatic essentialism completes a loop, from an interest in encoding the body into data, to an interest in programming and re-programming that genetic-informatic body, and finally to an investment in the capabilities of informatics to help synthesize and generate biological materiality.

Biotech is not about the re-affirmation of the body and materiality over and against the dematerializing tendencies of digital technology. Instead, it is about the mediation of this body-information episteme in a variety of concrete contexts criss-crossed by social, scientific, technological, and political lines. Biotech thus accomplishes this process through its tactical deployment of "biomedia" - the technical and pragmatic utilization of informatic essentialism towards the re-materialization of a range of biotechnical bodies.

What are the implications for this biotechnological investment in the body? For one, the fact of mediation is not being taken into consideration; the ways in which these various bio-technologies are not only intending to cure, but are significantly re-formulating what is meant by "body" and "health" are not under the main arena of consideration. "Health" - as a normative term - is never questioned in these contexts as to how it changes in different technological, political, and cultural instances. Along with this, the technologies in biotech are not simply objects or "things," but rather liminal techniques for intervening in the body; they operate not mechanically (as does a prosthetic), externally (as does surgery), or through engineered foreign elements (as does gene therapy), but by harnessing the "biological" (read: biological-as-natural) processes and directing them towards novel therapeutic ends. In such instances technology is indirect and facilitative; it is kept completely separated from the body of the (biomedical) subject: thus regenerative medicine's claim for a less technological, more natural approach to creating the context for advanced health.

In this way nature remains natural, the biological remains biological, plus, the natural and biological can now be altered without altering its essential properties, such as growth, replication, biochemistry, cellular metabolism, etc. The capacity of these technologies, and their aforementioned invisibility, enables researchers to conceive of a body that is not a body - a kind of lateral transcendence – a transcendence of abstraction. The technologies of therapeutic cloning, tissue engineering, and stem cell research, all point towards a notion of the body that is purified of undesirable elements (the markers of mortality, disease, instability, unpredictability), but that nevertheless still remains a body (a functioning organic-material substrate). The problems Hayles outlines - how to deal with the contingency of embodiedness - is here resolved through a revaluation and the production of a body purified through a combination of informatics and bioscience.

POST-ORGANIC LIFE

Posthumanism generally takes technological development as a key to the inevitable evolution of the human. However, it might be more accurate to call posthumanism a means of epistemologically managing the human and the technological domains. Posthumanism is, in a sense, an ambiguous form of humanism, inflected through advanced technologies. It seems that the posthuman wants it both ways: On the one hand, the posthuman invites the transformative capacities of new technologies; but on the other, it reserves the right for something called "the human" to somehow remain the same through those transformations. This contradiction enables posthuman thinkers to unproblematically claim a universality for attributes such as the faculty of reason, the inevitability of human evolution, or individual self-emergence. But many of the implications of posthuman technologies - distributed computing, computational biology, and intelligent systems - fundamentally challenge any position which places the human at its center.

But beyond this, what we find in contemporary biotechnology is a technically advanced, "thick" investment in the ways in which the body and information are directly related. Bioinformatics is perhaps unique because it is one of the few information sciences that is also a life science; its continued interest is not, then, in the anachronisms of the biological domain, but rather in the ways in which biology is itself a technology. Indeed, as science historian Robert Bud (1993) shows, the very meaning of the term "bio-technology" has, at least since the 19th century, indicated the industrial uses of naturally-occurring processes (such as fermentation, agriculture, livestock breeding) (see also Sunderland this volume). Contemporary molecular biotech follows in this tradition. Biotech is not to be confused with bioengineering or prosthetics; that is, biotech is not about interfacing the human with the machine, the organic with the non-organic. Rather, biotech is about a fundamental reconfiguration of the very processes which constitute the biological domain, and their use towards a range of ends, from new techniques in medicine, to new modes of agricultural production, to deterrence programs in biowarfare. As Bud states, biotech has always been about "the uses of life."

The culmination of these elements points to the fact that the condition for the future success of biotechnology will be the integration of information technology into the biological domain, while maintaining the ontological separation between human and computer under the ideology of the posthuman. In the biotech future, the body is approached as information, medicine becomes an issue of technical optimization, and "life" becomes an information science.

However, it would be too easy to fall into a position of either technophilia (where a more advanced biotechnology is the answer) or technophobia (where biotech carries the total burden of "dehumanization"). As one suggestion, we might look to those research endeavors within biotech which are adopting more sophisticated theoretical approaches to the intersections of bioscience and computer science, genetic and computer codes. For instance, research institutes such as the Biopathways Consortium and the Institute for Systems Biology are focusing not on the centrality of genes or DNA, but rather on biological "systems," biochemical pathways, and gene expression arrays.[iii] With a view to a systems-wide approach that would not reduce divergence or difference, one is reminded of Jorge Luis Borges' story, "The Garden of Forking Paths," or to the material uses of computer networks in communication. Similarly, unique collaborations between art and science in the domain of "bio art" are exploring the cultural, scientific, and political dimensions of fields such as cloning, New Reproductive Technologies, and connections between genetics and race.[iv] These are certainly not unproblematic approaches to thinking about the technoscientific body, and there is still much to be considered within research on the cultural valences of technoscience. But such examples may begin to demonstrate the ways in which technology is more than a tool, and that elusive materiality called the body is something other than the sum of its parts.

As genome projects are completed, genomic databases assembled, and as biotech becomes increasingly networked into mainstream health care, there thus needs to be a sustained, transformative intervention into the ways in which flesh is made into data, as well as the ways in which data is made flesh

NOTES

[i] For more see Howard (2000); Persidis (1999).

[ii] For more see textbook by Lanza et al., (1997)

[iii] For more see the Institute for Systems Biology's website at http://www.systemsbiology.org, and the group's proof-of-concept article, Ideker, T. et al., (2001).

[iv] Examples include the art collectives Biotechnika, Critical Art Ensemble, Biotech Hobbyist, SubRosa, and SymbioticA, as well as the art work of Susanne Anker, Brando Ballengée, Joe Davis, George Gessert, Eduardo Kac, Polona Tratnik, and Adam Zaretsky.

REFERENCES

Bud, R. (1993). *The uses of life: A history of biotechnology*. Cambridge: Cambridge University Press.

Ansell Pearson, K. (1997). *Viroid life*. New York: Routledge.

Bonassar, L., & Vacanti, J. (1998). Tissue engineering: The first decade and beyond. *Journal of Cellular Biochemistry, 30/31*, 297–303.

Halberstam, J., & Livingston, I. (Eds.). *Posthuman bodies*. Bloomington, IN: University of Indiana.

Hall, S. (2000, January 30). The recycled generation. *New York Times Magazine*, pp. 35–45.

Haraway, D. (1997). *Modest_witness*. New York: Routledge.

Haraway, D. (1991). *Simians, cyborgs, and women* (pp. 180–181). New York: Routledge.

Hayles, N. K. (1999). *How we became posthuman*. Chicago: University of Chicago.

Hood, L. *The human genome project and the future of biology*. Retrieved from http://www.biospace.com

Howard, K. (2000). The bioinformatics gold rush. *Scientific American*, July, 58–63.

Ideker, T., et al. (2001). Integrated genomic and proteomic analyses of a systematically perturbed metabolic network. *Science, 292*(May), 929–934.

Kurzweil, R. (1999). *The age of spiritual machines*. New York: Penguin.

Lanza, R., et al. (Eds.). (1997). *Principles in tissue engineering*. New York: Landes.

Latour, B. (1999). *Pandora's hope*. Cambridge: Harvard.

McLuhan, M. (1964/1995). *Understanding media*. Cambridge: MIT.

Mooney, D., & Antonios, M. (1999). Growing new organs. *Scientific American*, April, 60–67.

Moravec, H. (1998). *Mind children* (pp. 109–110). Cambridge: Cambridge University Press.

More, M. *The extropian principles: A transhumanist declaration*. Retrieved from http://www.extropy.org

Pedersen, R. (1999). Embryonic stem cells for medicine. *Scientific American*, April, 68–73.

Persidis, A. (1999). Bioinformatics. *Nature Biotechnology, 17*(August), 828–830.

Petit-Zeman, S. (2001). Regenerative medicine. *Nature Biotechnology, 19*(March), 201–206.

Shannon, C., & Weaver, W. (1965). *The mathematical theory of communication*. Chicago: University of Illinois.

Stocum, D. Regenerative biology & medicine in the 21st century. *E-biomed, 1*(March), 17–20.

Wiener, N. (1948/1996). *Cybernetics*. Cambridge: MIT.

Eugene Thacker
Literature, Communication, & Culture
Georgia Institute of Technology

ROSS BARNARD & DAMIAN HINE

4. BIOTECHNOLOGY HISTORY, ECONOMICS AND THE CARTESIAN DIVISION

Implications for the well-educated student

INTRODUCTION AND HISTORY

The average life expectancy at birth in western Europe in 1750 was 33 years. Today, in Australia, women live well into their 80s and men, on average, into the late 70s. A rapid increase in life expectancy began in the late 1790s, because of the discovery of vaccination (Jenner) and the discovery that 'germs' are the causative agents of disease. Significant contributors to these discoveries were Pasteur (Geison, 1995), Semmelweiss, Koch and Snow. Snow discovered the connection between contaminated water and disease (Vinten-Johansen et al., 2003). The two public health interventions that have had the greatest impact on the world's health are clean water and vaccines. Applying biological knowledge made it possible to intervene and improve the human condition through the prevention of disease. In the time of Jane Austen, a person over 40 years of age was considered to have one foot in the grave. Indeed until the 1940s, many women died before the age of 35, in large part due to the risks associated with childbirth. Many died young because of the unavailability of antibiotic treatment for infections, the lack of knowledge of antiseptic procedures, or the lack of knowledge regarding the need to wash hands between patients in hospital. Most of the deaths on the battlefield of World War One were from infected wounds. Relatively minor scratches are often lethal in an environment laden with *Clostridia* (anaerobic bacteria). We are all familiar with the great plagues of the middle ages that wiped out approximately one third of the population of Europe. The causative bacterium (*Yersinia pestis*) and the mode of transmission were not determined until the 1890s and the disease only became curable with the discovery of antibiotics tetracycline, streptomycin and gentamycin. Many of these advances were the product of a reductionist approach and the systematic analysis of the living organism as machine. Although this approach, is powerful in practice, it leads to an incomplete understanding of life. We contend that the graduate of the future needs to be familiar with the practical power of the systematic, reductionist approach, whilst being aware of the limits to knowledge imposed by the Cartesian division. By its definition (the *application* of biological knowledge to create products of value to society), biotechnology is based on a utilitarian approach and is product focussed. However, this product-focused research, which has transformed our lives, is based on decades of fundamental research. We need to ensure that graduates emerge with a deeper understanding of

Naomi Sunderland et al. (eds.), Towards Humane Technologies: Biotechnology, New Media and Ethics, 49–63.

scientific ontology and develop an understanding that the longer term health of a science-based economy will depend on basic research. The impoverished approach to research funding in Australia tends to lead to a short term focus that could be detrimental in the longer term.

Figure1. The plague doctor in 17th century Venice. The "beak" was stuffed with herbs and spices. Although the mode of transmission was not understood at the time, the waxed clothing may actually have helped prevent attachment of fleas and also reduce risk of droplet infection. Image used with permission of Omnia, Lido di Venezia, http://www.carnivalofvenice.com

DEFINITIONS

Definitions of biotechnology are contested. Various stakeholders posit and manipulate definitions to suit their own ends. An example of this type of manipulation is the classification and declassification (by both governments and the private sector) of industry sectors as belonging or not belonging to biotechnology on the basis of whether that classification is likely to appeal to potential investors or shareholders, or to enhance government statistics in that sector.

Our preference is to classify activities as biotechnology, based on the criterion of an affirmative answer to the question: does the activity apply biological knowledge for the generation of products that are, or *will be* valued by society? Value is contestable and changes over time. The things that a society values will depend upon the stage of development of the society and the risks to which it is currently exposed. Because starvation and disease are much greater risks in developing nations than they are in most parts of developed nations, biotechnology directed towards solving those problems will be more valued in those societies and the balance of risk versus gain will be very different. In Western nations since 1945, there has been a massive reduction in infectious disease burden and smallpox has been eradicated. People do not experience or witness the scourges (like polio) that were present until relatively recent times (even in R.B.'s own childhood of the late 1950s and early 60s), so their perceptions of risk and benefit have dramatically changed. There are complex philosophical issues surrounding the balance between public good (the utilitarian view of maximum benefit to the maximum number, versus individual rights and deontological ethical systems), but those debates tend to be easier to resolve in the face of epidemics, for example. Unemployment and economic downturn are risks that will unquestionably face developed economies if they do not move away from older industries, and invest in more innovation based industry (Hine et al., 2006; Quinlivan, 2006).

Many of our colleagues, in non-science *and* science fields have convenient or economically expedient definitions, fuzzy definitions, or do not have *any* consciously formulated definition. Some set up a definition that is framed in order to set up a "straw man", and, to the latter purpose, focus only on a narrow sector of biotechnology.

In such circumstances, it is best, in our view, to go back to basics and embed one's definition in the elementary, by considering the evolution of biotechnology. We will, first, put up some definitions that we have discovered, beginning with some fundamental definitions of the words, science and technology.

– Science is a process and set of ideas which have been constructed by people to explain everyday and unfamiliar phenomena (Layton, 1993; Kuhn, 1970). Science is conjectural, then subject to test (Popper, 1945).
– Technology is the scientific study of the practical or industrial arts (*The Oxford English Dictionary, 2nd edition*, 1989). Technology is a purposeful activity aimed at meeting needs and opportunities through developing products, systems and environments (Jones & Carr, 1993). Doing technology requires the fundamental endpoint: how well it fulfils its purpose in providing a solution to a problem (Layton, 1993).
– Bio: From the Greek, βιος, meaning "life, course or way of living" (*The Oxford English Dictionary, 2nd edition*, 1989).
– Technology is the employment of knowledge, hence biotechnology is the employment of knowledge about life.

The divisions between science and technology can be considered as societal. Communities that value knowing and communities that value doing. This is an artificial division. Real societies are a mix

We favour an evolutionary definition of biotechnology that puts biotechnology in historical context and captures the transition from ancient to modern. Of course the demarcation between traditional and modern biotechnology will be, by definition, dynamic and somewhat arbitrary.

This evolutionary definition also reminds the reader of the achievements of biotechnology, both in its traditional and modern forms. The modern reader does need to be reminded of these achievements, because the citizens of developed nations take an enormous amount for granted and make judgements from the perspective of rich and privileged beneficiaries of biotechnology.

The recasting of living things as being of potential human or commercial use is a *much* older practice than the modern biotechnology age (see, for example, Genesis chapter 1, verses 26 to 30). It did not just suddenly happen. It *can* be argued that in the modern age this recasting is more extensive and pervasive than it once was.

The history of biotechnology began when humans began to breed plants and animals, gather, test and process herbs for medicine; make bread and wine and beer; create fermented food products including yogurt, cheese and various soy products. Archeologists have unearthed examples of most of these processes that date back thousands of years. For example, preserved sprouting emmer wheat (*Triticum dicoccum* Schlübl) and dessicated yeast cells have been discovered at Amarna, dating back to the Egyptian New Kingdom (approx. 1550 to 1070 BCE). Samuel (1996) has beautiful scanning electron microscope images of the preserved yeast cells from pots in a workers village, Armana.

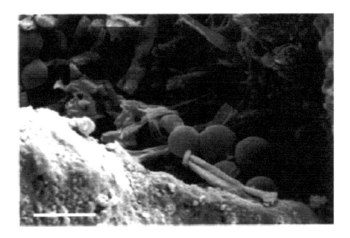

Figure. 2. Scanning electron micrograph of (dividing) yeast cells preserved in residue on a pot shard at Amarna workers vllage, Egypt. Image used with kind permission of Delwen Samuel, Kings College, London.

Methods of plant hybridisation and artificial selection (involving the random shuffling of genes through cross-breeding) have been extensively used over the centuries to change the genetics of plants to the point that most of our modern crop plants could not survive in the wild and are virtually unrecognisable compared to their wild ancestors. The modern potato and the modern tomato bear little physical resemblance to the forebears. Some 222 million acres world wide were planted with GM crops last year (2005). More than one-third was in developing countries. GM crops are presently grown by approximately 8.5 million farmers, with roughly 90 percent living in developing countries. That represents a massive contribution to the economies of the poorest countries in the world (Prakash, 2006).

The evolution of ancient biotechnology techniques into the modern biotechnology industry that uses recombinant DNA to create its products traverses milestones that includes smallpox vaccination (Jenner, 1796), the isolation of DNA by Miescher (circa 1871), rhabies vaccination (Pasteur, 1985), the discovery of penicillin by Fleming (1928) and its subsequent development by Florey, Chain and Heatley, the discovery of the structure of DNA in 1953 by Watson, Crick, Franklin and Wilkins, the deciphering of the genetic code in 1961 by Nirenberg and Khorana, polio vaccination (Salk and Sabin) the first recombinant DNA experiments in 1973 by Gilbert, the creation of the first hybridomas in 1975 (Kohler & Milstein), the founding of Genentech Inc. in 1976 (Swanson & Boyer) , the production of the first monoclonal antibodies for diagnostics in 1982 and the production of the first recombinant human therapeutic protein (humulin) in 1982.

Since the manufacture of human insulin using recombinant *Escherichia coli* in 1982, many human therapeutic or vaccine proteins made by modern biotechnology methods have now been approved for marketing after extensive pre-clinical and clinical trialling, by the FDA (US Food and Drug Administration). Wallman, (1997) gives a (nonexhaustive) list, with date of approval by the FDA: Protropin (1985), Intron A (1986), RECOMBIVAX HB (1986), Roferon-A (1986), Humulin (1987), Humatrope (1987), Alferon N (1989), Engerix-B (1989), Actimmune (1990), Activase (this is human, recombinant tPA) (1990), Leukine (1991), RECOMBINATE (1992), Proleukin (1992), OncoScint (1992), Pulmozyme (1993), Betaseron (1993), EPOGEN (1993), PROCRIT (1993), KoGENate (1993), ORTHOCLONE (1993), NEUPOGEN (1994), Nutropin (1994), ReoPro (1994), Cerezyme (1994), Rituximab (1997) and Herceptin (1998). The last two listed are treatments for non-Hodgkin's lymphoma and breast cancer, respectively.

There are now several hundred human therapeutic drugs, gene therapies and vaccines in clinical trials. Some of these are produced by recombinant DNA technology, some are purified natural products, some are chemically synthesised and some are chemically sythesised modifications of natural products. Products are being tested to target the following diseases: cancer, AIDS, heart disease, multiple sclerosis, rheumatoid arthritis and viral diseases. Products are being developed to reduce bleeding from surgical procedures, accelerate wound healing and prevent organ transplant rejection. New and effective combination drug therapies have been developed against AIDS and Hepatitis. The anti-breast cancer drug, *Herceptin*, a recombinant, humanised, mouse monoclonal antibody that kills cancer cells, has recently become widely available in Australia. Research is under way to produce a vaccine against malaria and to improve mosquito control programs (some of the latter work, funded by the Gates foundation, is being carried out by researchers and biotechnology students at the University of Queensland). A recombinant vaccine to protect against cervical cancer (also developed at the University of Queensland) has been successful in clinical trials and has just been launched. The human genome project, and the genome projects of multiple other species (bovine, murine, rice genome projects, for example) have provided enormous databases of genetic information. This has radically changed the paradigm for drug discovery (see Fig. 2). It will facilitate discovery of molecular targets for new therapeutic compounds, as well as the discovery of disease markers for diagnostic purposes.

The issue of access to new technologies is an enormous, complex and thorny one. It is an inevitable fact that most new technologies will be developed in the rich nations and that the technologies will, in general, be those that can generate income for the companies developing them. This is particularly the case for development of new pharmaceuticals, where the development and clinical testing (mandated by law and regulated by the TGA or FDA) takes typically 15 years, and the costs of development and clinical trial must be recovered in approximately 5 years.

It will be crucial that there be parallel improvements in public health infrastructure, transportation, food and medicine storage facilities, if benefits are to flow to the developing nations. It is one thing to make vaccines available, it is another to ensure that the *cold chain* is not broken during transportation and storage. It is essential to increase crop yields and nutrient content through application of technology (of any kind), but the effort will be wasted if we do not prevent access of rodents to grain storage facilities.

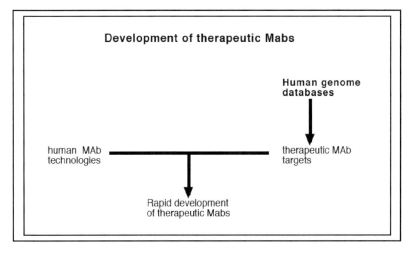

Figure 3. The new paradigm, made possible by the human genome project, for development of therapeutic monoclonal antibody (MAb) products.

One of the major themes of this book is the mediations of biotechnology. In relation to causes of the predominant mediations of biotechnology, immediately one invokes applications and *employment* of knowledge, one enters the realm of economics and politics. Any economic transaction requires the definition of one person (or group) as a buyer and another as a seller. Perhaps the alienating use of language is one inevitable consequence of our economic system? Key questions are, *why* are the current mediations of biotechnology adopted as a matter of public policy at state, national and international levels and why should they be so adopted? Do the mediations match the reality of government spending priorities?

ECONOMIC CHANGE AND IMPLICATIONS FOR EDUCATION AND TRAINING

Science and engineering are at the heart of the 21st century. New knowledge is a powerful driver of economic prosperity and a force for human progress. That makes new knowledge the most sought after prize in the world". (Dr Rita R. Colwell, Director of US National Science Foundation 1998 – 2004)

Much of what we say in this section is premised on the assumption that it is desirable for Australians to maintain or improve their living standard, including their standard of education, health and disposable income over the ensuing decades. One can argue over the distribution of wealth within Australian society, and the mechanisms for achieving that distribution, but if we do not have wealth to distribute, then that argument becomes theoretical.

Schumpeter's fundamental, frequently cited, early work on technological processes and economic development (Schumpeter, 1934) has led to the widely accepted view that the long term performance of the economy of a society is dependent on the ability of the people to innovate and exploit new technologies. Subsequent work supports the view that a country's gross domestic product is substantially enhanced by increases in scientific research and technological developments. One such study attributed half of the increase in US GDP in the 20th century to the exploitation of new knowledge (The Council of Economic Advisers, 1995).

In industrialized societies, consumers constitute an increasing majority over producers (due to mechanization of production and concentration of ownership of the means of production). This is illustrated when one considers the changing demographics and efficiency of agricultural production in the US over time. In 1960, the US produced 252 million tons of food. By 2000, this had increased to 650 million tons of food on 25 million fewer acres of land. In 1940 each US farmer fed 19 people. Today each farmer feeds 140 people. The number of farmers has reduced, whilst the crop yield per hectare and per farmer has inceased astonishingly (Prakash, 2006).

Because consumers are in an increasing majority, popularly elected governments will, inevitably, be preoccupied with influencing consumer perceptions.

Despite the statements by Australian governments about support for innovative industry and building long term economic strength through investment in science and technology and science and technology education, this is not, unfortunately, matched by reality (Hine et al. 2006). Furthermore, at federal level, government is not pulling its weight in terms of investment in basic research. This creates a condition where alliances with industry are necessary to continue research. However, the investment by Australian industry in R & D compares very poorly with the international situation.

Governments at federal and state level have given some support for the growth of knowledge-based industries through initiatives such as those recently developed in Queensland (the Smart State initiative) and Victoria (Bio 21). These initiatives will not be sustainable if they are not supported by a serious commitment to education. In 2005 the Royal Australian Chemical Institute published the results of a survey of employers on their success or otherwise in recruiting qualified staff. For most forms of chemistry, a high percentage of employers reported difficulty in recruiting qualified staff. In biotechnology the figure was as high as 79%, and in the newest field of industrial endeavour, nanotechnology (a field at the intersection of polymer chemistry, interfacial chemistry, physics and biology) the figure rose to 82% (RACI, 2005). The RACI survey is consistent with the findings of the DEST/AUTC report on biotechnology education in Australia (Gray et al., 2003). This survey reported results from detailed responses by more than 60 biotechnology companies. It identified that the greatest demand and greatest challenge in recruiting for the Australian biotechnology industry was to find graduates in chemistry and molecular biology. It also identified those areas as the ones that will be most affected by technological developments. It is very important to note that, after technical skill, prospective employers rated the following attributes very highly: enthusiasm and willingness to learn, problem solving, critical thinking and creativity, interpersonal skills and teamwork. These responses were consistent across different sectors of the biotechnology industry and were consistent in responses to both open-ended questions and closed scale questions.

A globally mobile workforce is already upon us, and a major challenge exists in retaining our graduates who will be attracted by nations which *are* investing in innovative industries, like biotechnology. At the present time more than 10% of our 18-to-35-year-olds live overseas, and most of these are highly educated and will work for foreign economies during their most productive years (Hine et al, 2006).

An international conference on science teaching and research held at the University of Queensland (Mattick & McManus, 2004) highlighted the severe strain that tertiary science education in Australia is currently experiencing. Enrolments in the enabling sciences, chemistry, physics and mathematics, have declined by 23.2%, from 17,064 in 1989 to 13,105 in 2002. The longer term prospects of turning this around are not promising, because, as an Australian Council of Deans of Science (2002) study has highlighted, the decreasing science enrolments at high school go a long way to explaining the decrease in science enrolments at University.

It can be quite reasonably argued that the quality of the educational experience of a university science student has diminished. Science faculties are reducing practical class teaching to cut costs and student teacher ratios are unacceptably high compared to other OECD countries. Entry scores for science and technology degrees have decreased substantially over the last 5 years as students migrate to professionally orientated degrees such as medicine, pharmacy, engineering or commerce. This should not be surprising in view of the increasing cost of tertiary education in a "user pays" environment. This could result in a situation where we do not have the raw material to produce top-tier scientists and top-tier, basic research (as defined by international impact).

The commitment of the Australian business community to supporting tertiary graduates and investing in research and development is crucial. Investment in R&D is an investment in the future products, innovations and development of the entire economy. The federal government investment in R&D is comparable to other OECD countries, but Australia's ranking is falling. In 1996/7 Australia was ranked 3rd amongst OECD countries for GERD and by 2002/3 this had slipped to 7th. The contribution of the States, does not come close to that of the Federal spending, as would be expected given much of the R&D infrastructure (CSIRO and Universities) is largely federally funded. Unfortunately most states have shown a marked decline in R&D expenditure in recent years (SA and QLD are notable exceptions), while the Federal Government's expenditure is stable (ABS Statistics, 2005).

It should be of great concern that investment at both the federal and state government levels is remaining static or declining in the context of the growing international competition in science and technology. If one considers China and the US over the period 1990 to 2003, the former tripled the percentage of its high-tech manufacturing from 6% to 18% of total output, while the United States raised its output from 12% to 30% of total. If one considers U.S. patents as an indicator of scientific health, then the growth in our Asian neighbours, South Korea, Indonesia, India, Malaysia, Philippines, Singapore, Taiwan, Thailand and China (Asia-8) accounted for 4.2% and 18.4% of all foreign patent applicants in 1990 and 2003, respectively. Over the same timeframe Australia accounted for 1.1% and 1.5%, respectively. Further, the collective share of published scientific articles by the Asia-8 has risen from less than 4 percent of the world total in 1988 to 10 percent in 2003. Australia has only increased from 2.3 to 2.9 % over the same period (Hine et al., 2006). These data indicate that our regional neighbours are focusing on developing knowledge-led economies and will be fierce competitors not only in high volume manufacturing, but also in an innovative space we need to enter in order to maintain living standards.

The bulk of R&D expenditure in many OECD countries is carried out by the private sector. However, in this area, Australia ranks 16th, well below countries with similar population and living standards, such as Singapore and Canada. The investment in R&D by leading Australian companies pales into insignificance against their international counterparts. Expenditure on R&D as a percentage of total assets by our life sciences leader, CSL for 2004 was 2.6%, while Genentech spent 10.1%. In mining, BHP spent 0.04% of total assets on R&D, while Newmont Mines from the US spent 1.5%; and in petroleum, Woodside Petroleum spent 0.04%, while Total of France spent 0.8%. In all our comparisons of sector leaders, Australian companies fared poorly, even taking into account our noted productivity. With such low spending directed at developing new products and services, the engine of a company's growth, it is difficult to see how Australia can or will do anything to reverse the long decline in its industrial competitiveness and relative GDP.

Unfortunately, at the moment we have an ominous combination of decreasing government expenditure on research and development and business expenditure on research and development compared to other economies, plus a flight from science at both the high school and university level. The fact this has happened when we are experiencing unprecedented economic growth, reflected in large Federal government surpluses and significant gains in GST revenue for State governments, a time of opportunity for investment in the future, severely compromises our future competitiveness.

A "heads up" to the US government was provided by a National Academy of Sciences report entitled *Rising above the Gathering Storm: Energising and Employing America for a brighter Economic Future.* This report made four recommendations: (1) Increase America's talent pool by vastly improving K-12 science and mathematics education; (2) sustain and strengthen the nation's traditional commitment to long-term basic research that has the potential to be transformational to maintain the flow of new ideas that fuel the economy, provide security, and enhance the quality of life; (3) make the United States the most attractive setting in which to study and perform research so that we can develop, recruit, and retain the best and brightest students, scientists, and engineers from within the United States and throughout the world and; (4) ensure that the United States is the premier place in the world to innovate; invest in downstream activities such as manufacturing and marketing; and create high-paying jobs that are based on innovation by modernizing the patent system, realigning tax policies to encourage innovation, and ensuring affordable broadband access.

The preceding paragraph highlights that there is international recognition of the need for a strong foundation upon which to build a knowledge centred economy and the resolution with which the US is likely to strive to attract the best students and scientists in the near future. It requires a vision that spans from primary and secondary schooling, to university and beyond. It demands a better understanding between the different sectors. In the past, governments have been willing to fund higher education to advance the frontiers of science and to produce an educated community. Unfortunately, recent Australian governments with, typically, a two and a half year vision, appear to have little interest in the huge longer term economic benefit of investment in research. Yet, if economists insist, the economic benefits of medical research can be quantified. Medical research budget in the U.S. was 36 billion dollars in 1995. If 10% of the increased human longevity (average approx. 6 years) was attributable to medical research (allowing for the fact that much of the increase is related to higher incomes, better diet, less smoking etc.) the economic return amounts to 240 billion per year (Murphy & Topel, 2000).

The face of tertiary education in Australia has changed dramatically. Students are paying increasing amounts of money for education, either through the HECS scheme or upfront fees. There has been a parallel change in student expectations regarding outcomes and of the purpose of education. Students are now customers. They are pragmatic. In our direct experience with biotechnology students over the last six years, they expect employment, career path and appropriate financial reward. Will students be adequately trained to take up positions in the emerging industries? Will there be sufficient numbers of these graduates to fuel the new innovative economy? Will there be sufficient investment in basic research to sustain innovation over the longer term? Will there be investment in core training in fundamental skills like chemistry. Not unless there is a change in the course of government policy and emergence of a vision beyond the electoral cycle.

THE SCHISM IN WORLD VIEWS: A CHALLENGE FOR TRANSDISCIPLINARY EDUCATION

Albert Einstein (1954) captured this schism elegantly:

> Now it is characteristic of thought in physics, as of thought in natural science generally, that it endeavours in principle to make do with "space-like" concepts *alone*, and strives to express with their aid all relations having the forms of laws. The physicist seeks to reduce colours and tones to vibrations; the physiologist, thought and pain to nerve processes, in such a way that the psychical element as such is eliminated from the causal nexus of existence, and thus nowhere occurs as an independent link in the causal associations.

There is a related schism between the world views and the ways of understanding promulgated in our schools of science and our schools of humanities. It was asserted by Barnard & Turnbull (2006), that neither view is complete and that there is a need for a transdisciplinary *third space* for dialogue and an emphasis on this third space in our educational work programs.

We assert that an appropriate educational paradigm is one that a) maintains core technical training and competence in the practical and applied aspects of science, but, b) encompasses this within a broader epistemological understanding of the nature of knowledge and how new knowledge is generated (a conjectural process, requiring leaps of imagination, modelling and prediction, followed by testing (Popper, 1945) and the indispensible role of fundamental research. Science is, at its core, the human imagination constructing models, followed by testing to see how closely these models approximate reality and how well they predict what we see in nature. Heisenberg (1959) has a lot to teach, and his work should be used as a teaching resource for courses in the fundamentals of science.

It is true that the Cartesian division (and the alienation of mind from body, the perception of the body as object/machine) has lot to answer for, but the approach has had many positive outcomes in terms of the advancement of our understanding of the *mechanisms and causative agents* of human disease and understanding of the functioning of living things. It is also pertinent to recall the poor treatment of the mentally ill as recently as the 20th century. Only with increasing understanding of the physical bases of mental illness was stigma and individual guilt alleviated and treatment of these people improved (Barnard & Turnbull, 2006). These are advances that would have been impossible without the objectification of living things (including humans).

It is important to note this "objectification" of living things is *not* a new phenomenon, but is a view that stems back to the renaissance and post-renaissance scholars (Descartes, Huygens, Hooke, Leibnitz, Galileo, Leeuwenhoek, DaVinci, William Harvey, Lavoisier). It could be argued that the process in biotechnology simply continues this tradition, albeit with an (arguably) stronger politico-economic dimension (although the political dimension was strong in the time of

Galileo, and it was a Cartesian view, a separation between the jurisdictions of science and religion, that allowed some scientists to continue working in the shadow of a dogmatic Church).

The formulation of questions as a schism between opposites is a tradition that is continued in the field of textual analysis and in debate. Derrida (1988) maintained that text usually depends on binary oppositions; something is (a) and therefore not (b). Deconstruction first aims to bring out these oppositions and then to displace them by pluralism. In this way, it allows the reader to see (a) as something other than not (b). The process of deconstruction also allows examining what is left out, such as topics, authors, ideas, and deviations from the premise(s) of the text. In the debate about parts of biotechnology, in many cases in my experience, the first stage is undertaken, but the examination of what is left out, the areas that do not fit neatly into preconstructed categories, that constitute a *tested* reality (see Popper, 1945) independent of our preconceived notions, receive scant attention.

Much debate in the area surrounding biotechnology, in the realms of ethics, GM foods, stem cells, nanotechnology, assisted reproductive technology is couched in "black and white", catch-all oppositions. Much critical analysis of biotechnology is based on an assertion of a value system that entails classification of particular things as "good" and other things as "bad". The fact that a choice of values has been made is often not explicitly acknowledged "up-front" but it should be. The axioms of good and evil often owe much to a morality of inclusion (of the type discussed by Buchanan et al. (2000)). Discussion of moral frameworks is important, because people make assertions based on different value systems. For example, they might begin with the value based premise that market forces are "good" whereas government intervention is "bad", or *vice versa*. If this premise is not made explicit then this is tantamount to an implication that there is one shared value system amongst the audience or the readership, a concept quite contradictory to the idea of heteroglossy and, we think, lacking in respect for the diverse of value systems in the audience.

The dualisms of Plumwood (1993), for example, are "black and white" constructs that need to be justified and many examples are sweeping generalizations that are used to support arguments but are not founded on a deeper and a more than superficial understanding of biotechnology and biotechnologists. Let's examine the assertion by Plumwood:

> Hence the speed of direct genetic manipulation at the molecular level is esteemed over the longer time periods associated with traditional cross breeding.

In fact *many* plant biotechnologists, including Howarth Bouis, assert that *both* cross breeding *and* molecular technologies should be used, because each technology has practical and economic weaknesses and advantages in particular circumstances (Bouis, 2000). Speed *is* esteemed in some cases in plant biotechnology and in many areas of biotechnology (especially in genome sequencing and in pharmaceutical discovery and development, because of the highly competitive nature of these fields and the very limited time available to recoup research and clinical trial costs).

Another example is the generalisation that genetically modified crops are destructive of traditional agricultural competencies. Some modern plant bio-technologies are *not* competency destroying (eg. insect resistant crops, herbicide tolerant crops, drought tolerant crops, submersion tolerant rice, golden rice, synchronously ripening crops). The same or very similar sowing and tillage practices can be employed, even though less tillage may be required and less pesticide need be applied. However, some GM crops (like those incorporating terminator technology) have the potential to be competency destroying. What is beyond doubt is the massive (of the order of 70%) reduction in application of organochlorine or organophosphorous pesticides and increase in cotton crop yields in India (Qaim and Zilberman, 2003) and Australia (Doyle et al., 2002) since the introduction of GM crops, which is a tremendous positive for the environment.

The absolutism of the categorization by some authors is quite frightening to us. We assert that for reasoned debate we need to displace "black and white" oppositions by pluralism and a more detailed dissection of technologies and issues on a case-by-case basis.

Figure 4. The cathedral at Chartres, constructed between 1194 and 1260. ©Artifice, Inc./GreatBuildings.com

With a reasoned dialogue, transformation of biotechnology practice may be possible. A well rounded student, who combines technical and scientific skill with creativity and communication skills, might be better able to participate in such dialogues. They might be better able to pose difficult questions and make informed suggestions about the directions of scientific research and development. Is it desirable for these directions to be shaped by unfettered market demand, as opposed to having ethical principles determine priorities (eg. channeling funds to cure devastating diseases that afflict large numbers of people-rather than to the enhancement of normal traits or discovery of treatments for male baldness)

(Buchanan et al., 2000)? Or, (in the pluralist view) should *both* of these forces operate to drive research and development? Market forces are acknowledged by many biotechnologists and some economists as an imperfect driver of research in some areas, but very good in others. The questions then become ones of balance of these "black and white" alternatives and of government priorities in resource allocation which is, unfortunately, conditioned by the short term whims of the electorate.

We will end by returning to the middle ages. It has been argued that the renaissance emerged on the back of a *transdisciplinary synthesis* of technological and artistic expertise that was developed for the building of the great cathedrals. In these massive constructions, architectural vision was systematically documented for the first time. Mathematical, technical and artistic skill was brought to bear to realize an object of symmetry and beauty that had a *social purpose*. Perhaps in the choice of directions for biotechnology education we should be aiming to produce graduates who are capable of achieving an analogous synthesis. That is, capable of mediating a dialogue between value systems to build a structure with a social purpose. This will require graduate attributes that are, perhaps paradoxically, consistent with the expressed demands of the biotechnology industry

ACKNOWLEDGMENTS

Thank you to Maureen Cavanagh for critical reading and bringing some details of the history of the plague doctor to our attention. This work was partly funded by the Carrick Institute for Learning and Teaching in Higher Education

REFERENCES

Australian Bureau of Statistics. (2005). Retrieved from http://www.abs.gov.au/
Australian Council of Deans of Science. (2002, June 24). *Higher education review submission 38.*
Barnard, R. T., & Turnbull, D. (2006). On the incompatibility of genetic axioms with axioms of justice and ethics. *The International Journal of Interdisciplinary Social Sciences, 1* (in press).
Bouis, H. (2000). In M. Qaim, A. Krattiger, & J. von Braun (Eds.), *Agricultural biotechnology in developing countries: Towards optimizing the benefits for the poor* (Chap. 11). Kluwer Academic Publishers.
Buchanan, A., Brock, D. W., Daniels, N., & Wikler, D. (2000). *From chance to choice.* Cambridge University Press.
Council of Economic Advisors. (1995). *Supporting research and development to promote economic growth: The federal government's role.* Washington, DC: Executive Office of the President.
Derrida, J. (1988). *Limited Inc.* Evanston, IL: Northwestern University Press.
Doyle, B., Reeve, I., & Barclay, E. (2002). *The performance of ingard cotton in Australia during the 2000/2001 season.* Report by Institute for Rural Futures, University of New England.
Einstein, A. (1953/2004). *Ideas and opinions. 19th impression.* New Delhi: Rupa & Co.
Geison, G. L. (1995). *The private science of Louis Pasteur.* Princeton, NJ: Princeton University Press.
Gray, P., Barnard, R., Franco, C., Rifkin, W., Hine, D., & Young, F. (2003). *Review of Australian bio-technology education.* Department of Education, Science and Training and the Australian Universities Teaching Committee. Commonwealth of Australia.
Hine, D., Mattick, L., Barnard, R., & McManus, M. E. (2006). Is global S & T built on a house of cards? *Australasian Science, 27*(8), 37–40.
Heisenberg, W. (1959). *Physics and philosophy: The revolution in modern science.* London: Allen & Unwin.

Jones & Carr. (1993). Teachers' perceptions of technology education. In *Towards technology education*. Hamilton: Centre for Science and Mathematical Education Research, University of Waikato.

Kuhn, T. S. (1970). *The structure of scientific revolutions*. Chicago: University of Chicago Press.

Layton. (1993). *Technology's challenge to science education*. Buckingham: Open University Press.

Mattick, L. E., & McManus, M. E. (2004). *Science teaching and research: Which way forward for Australian Universities*. Conference report. Retrieved from http://www.brightminds.uq.edu.au/TRC/report.htm

Murphy, K., & Topel, R. (2000, May). *Exceptional returns: The economic value of America's investment in medical research*. Mary Woodward Lasker Charitable Trust.

National Academy of Sciences, Committee on Science, Engineering and Public Policy. (2006). *Rising above the gathering storm: Energizing and employing America for a brighter economic future*. Washington, DC: The National Academies Press.

Popper, K. (1945). *The open society and its enemies*. London: Routledge and Sons.

Plumwood, V. (1993). *Feminism and the mastery of nature*. London: Routledge.

Prakash, C. S. (2006). *Academic and science community applauds WTO GMO ruling*. http://www.agbioworld.org/biotech-info/pr/wtoruling.html

Qaim, M., & Zilberman, D. (2003). Yield effects of genetically modified crops in developing countries. *Science, 299*, 900–902.

Quinlivan, B. (2006, April 27–May 3). High-tech laggards. *Business Review Weekly*.

Royal Australian Chemical Institute. (2005). *Future of chemistry study: Supply and demand of chemists*. Final report. RACI.

Samuel, D. (1996). Archaelogy of ancient Egyptian beer. *Journal of the American Society of Brewing Chemists, 54*(1), 3–12.

Schumpeter, J. A. (1934). *The theory of economic development: An inquiry into profits, capital, credit, interest, and the business cycle* (R. Opie (Trans.). Cambridge, MA: Harvard University Press.

Vinten-Johansen, P., Brody, H., Paneth, N., Rachman, S., & Rip, M. (2003). *Cholera, chloroform, and the science of medicine: A life of John Snow*. Oxford University Press.

Wallman, S. (1997). *A short history of biotechnology*. New Hampshire Community Technical College. http://64.72.11.16/biotech/

Ross Barnard,
Biotechnology Program
School of Molecular and Microbial Sciences
University of Queensland,

Damian Hine
Business School
University of Queensland

JOSEPH HENRY VOGEL

5. NOTHING IN BIOPROSPECTING MAKES SENSE[i]
EXCPECT IN THE LIGHT OF ECONOMICS

Economics is a theoretical structure which can make a meaningful picture of a pile
of sundry facts on bioprospecting. To date, the media has reported bioprospecting
sans economics and the result has been counterproductive to the spirit and letter of
the Convention on Biological Diversity. An example from *The New York Times* is
analyzed which, in the light of economics, would justify a biodiversity cartel.
Similarly, in the light of economics, "The Bonn Guidelines on Access to Genetic
Resources and the Fair and Equitable Sharing of Benefits Arising out of their
Utilization" are little more than an instrument for biofraud.

I. INTRODUCTION

When Theodosius Dobzhansky wrote "nothing in biology makes sense except in
the light of evolution" (1973), he was responding to the creationist nonsense of the
day. A generation later, biology teachers will still cite Dobzhansky as if to say "we
are not going to waste precious class time debating creationism." Something similar
can now be said with respect to the biotechnologies that derive from genetic
resources: "Nothing in bioprospecting makes sense except in the light of economics."
As I shall develop in the next few pages, economic theory can explain why benefit-
sharing did not pan out as originally foreseen and, much more importantly, what
can now be done. On its own merits, the analysis to follow will probably not make
it into the public discourse. This means that the question before us is mundane:
How do we engage civil society to mount pressure on politicians to reform policy
in the light of economics? The answer lies in the art of communication where
economists do not hold the comparative advantage. We tend to write in a dense
language and for a select few. In contrast, the journalist is a generalist who writes
for the masses and measures his or her success by how well disparate events are
integrated into a flowing narrative. As long as the economist can provide an integrating structure for that narrative, an opportunity arises for trade. This chapter will
explain how journalists can make sense of bioprospecting in the light of economics
and, along the way, suggest a few stories that have not yet been adequately told.

II. REPORTING *SANS* ECONOMICS

To make economic sense of bioprospecting will require a certain degree of mental
discipline that goes beyond *status quo* reporting. Because thinking like an economist

Naomi Sunderland et al. (eds.), Towards Humane Technologies: Biotechnology,
New Media and Ethics, 65–73.

does not come easily, even for economists, many journalists will be reluctant. Applying abstract theory is work and journalists may fear that the very language of economics will turn off editors and readers alike. Fortunately, the economics needed to make sense of bioprospecting is not *that* difficult and corresponds to what is covered in an introductory college course. Given the fact that tens of millions of people around the world have formally studied economics, and a good number are concerned with biodiversity conservation, the potential readership is huge. One could even say that the serious journalist has little choice but to report bioprospecting in the light of economics. Adapting another apt phrase from Dobzhansky's seminal article: "Without that light, [bioprospecting] becomes a pile of sundry facts some of them interesting or curious but making no meaningful picture as a whole". (Box 1)

Box 1. Possible Stories on Bioprospecting in the Light of Economics

1.	Talk is NOT cheap: A running tab on the Conferences to the Parties of the Convention on Biological Diversity (COPI-COPIX)
2.	Is the "Group of Like-Minded Megadiverse Countries" a Biodiversity Cartel?
3.	Governments selling genetic resources and not disclosing the price: Welcome to the commercial secrecy of bioprospecting!
4.	More than fifteen years of the Convention on Biological Diversity…no let up in the mass extinction crisis
5.	How much is that frog worth? Who owns it? *Epipedobates tricolor* and biopiracy
6.	One hundred ninety countries on board to the Convention on Biological Diversitiy---Notable exceptions: the U.S. and Iraq
7.	Despite much publicity, can Costa Rica be a model for bioprospecting?
8.	Privatizing benefits and socializing costs: the contentious economics of R&D
9.	Biopiracy morphs into biofraud: The Bonn Guidelines to the Convention on Biological Diversity
10.	Gene-culture co-evolution: A bugaboo in bioprospecting deals?

Nothing persuades like a well-chosen example. So, I have combed through the news media to find an article that epitomizes both the problem of reporting bioprospecting *sans* economics as well as the opportunity of making whole the sundry facts presented. I have found a fairly comprehensive story (2090 words) entitled "Biologists Sought a Treaty; Now they Fault it" which appeared on the May 7, 2002 edition of the front page of the Science Times Section of *The New York Times*, approximately one month after the sixth Conference of the Parties to the Convention on Biological Diversity (COPVI). The author is Andrew Revkin who has won accolades in journalism for going the extra mile to get the story right (quite literally in the case of *The Burning Season*). Despite Revkin's scientific background and journalistic skill, there is no economic reasoning in "Biologists Sought a Treaty; Now they Fault it." This should not surprise us. The sheer volume of Revkin's productivity (some 979 bylines in the NYT since 1981) may preclude the downtime necessary for making economic sense of bioprospecting. Moreover, the work in organizing the story in the light of economics is intrinsically different than that of tracking down facts and tying them up into crisp prose. An economic interpretation requires that the journalist tease out causes and effects and then display the chutzpah to voice unwelcome implications. The result will be a bioprospecting story that will not sit well with many of the sources of the sundry facts. Inasmuch as Revkin has not done this, an opportunity arises to unify and connect the hitherto unrelated facts presented in "Biologists Sought a Treaty; Now they Fault it."

Irony carries a certain *cachet* and Revkin sets the tone beginning with the title. Pity that it misrepresents the sequence of events that culminated in The Convention on Biological Diversity. The text of the Convention was the product of arduous negotiations that took place in the late 1980s and early 90s under the auspices of the UNEP in Nairobi, Kenya. The representatives from the North and South were so divided that they never settled their differences; instead they immortalized them in a wishy-washy language that was faxed out of Africa just hours before the inauguration of The Earth Summit in Rio de Janeiro, in June 1992. Revkin's article could have just as ironically, and much more accurately, been entitled "Critics of The Convention on Biological Diversity foresaw its Failure from the Get-Go". (Figure 1)

Figure1. Fora Bush. Citing concerns over intellectual property rights, Bush-père refused to sign the Convention on Biological Diversity at the Earth Summit, Rio de Janeiro, June 1992. The Portuguese banner reads "Bush, Go Home!... The Amazon Is Ours!..." In the new millennium, little has changed.

The misconstrued irony in Revkin's title is compounded by another in the first 100 words of the article "…biologists say, in many tropical regions it is easier to cut a forest than to study it." That salvo is supported by a quote from Dr. Douglas C. Daly, curator at the New York Botanical Gardens, "'Something that was well intentioned and needed has been taken to an illogical extreme.'" Revkin then builds the story to elucidate Daly's thesis, showing how nationalist groups have become so obsessed with biopiracy that they are running scientists not only out of countries but out of their own fields of study. "Christiane Ehringhaus, a German botanist pursuing a doctorate at Yale, was teaching Brazilian students and studying plants in the state of Acre in the Brazilian Amazon when newspapers implied that she was collecting seeds and insights from indigenous people in pursuit of drugs… the resulting difficulties had prompted her to abandon botany altogether…"

III. REPORTING IN THE LIGHT OF ECONOMICS

With a modicum of economic theory and the chutzpah to expose the vested interests in bioprospecting, Revkin could have integrated the sundry facts into a more comprehensible whole. Let's begin with that opening statement about the relative ease of cutting the trees versus the difficulty in studying them and then move on to Christiane Ehringhaus. The essential fact needed to resolve the paradox is this: the U.S. has not ratified the Convention on Biological Diversity and so the Convention is not binding on the U.S. Revkin missed the significance of this fact: genetic resources that wind up in the U.S., by hook or by crook, are the property of no one (*res nullius*) and hence fair game for Research and Development (R&D). Any biologist who comes to study the environment may inadvertently (or advertently) facilitate biopiracy. Regarding the disheartened Christiane, Revkin quotes her as saying: "First…they drove me completely away from medicinal plants and now from plants, period." That comment should have raised a red flag. Any botanist researching medicinal plants is there to throw such knowledge into the public domain through publication, thereby disenfranchising both the country of origin and the traditional peoples. Should the published traditional knowledge provide a lead for R&D, then the resultant biotechnology will enjoy a monopoly patent both in the U.S. and, under the Trade Related Intellectual Property Rights (TRIPS), the country of origin. What Revkin portrays as xenophobia---the exclusion of foreigners from collecting medicinal plants in Brazil---is economically quite sensible. Inasmuch as there has also been gene-culture co-evolution between ancestral peoples and their environment, the exclusion of all plants makes similar economic sense.

A caveat is in order: locking up access to genetic resources is not without its costs. No doubt taxonomy is being thwarted through exclusion; without taxonomy, conservation cannot proceed. In response to such complaints, the Brazilian government in October 2003, eliminated much of the red tape for public science. This was a calculated decision that also makes sense in the light of economics. The probability that a valuable biotechnology will emerge serendipitously from a taxonomic specimen is extremely slim. One need only note that the much-heralded INBio (Instituto Nacional de Biodiversidad) of Costa Rica has had no blockbuster products after more than ten years of intensive screenings. However, a slim chance is not a zero probability. The case of *John Moore v The Regents of California* illustrates how samples collected for public science (John donated his spleen) can be appropriated for private profit and result in a multi-billion dollar industry (interleukin and interferon). Should some biotech company exploit the loophole that all specimens in the U.S. are *res nullius*, then a future Brazilian government will rue the day that they gave the taxonomists a pass.

Damned if you grant access to genetic resources and damned if you don't: what then is the economic approach to bioprospecting? The answer will depend on the economist you talk to! Those who believe that economics is a science will try to put a value on biodiversity in its myriad of uses, both direct and indirect (see, for example, Munasinghe, 1992). Once those values are quantified and aggregated, then the economist-as-scientist will boldly proceed to determine the optimal level of habitat conservation through cost-benefit analysis. With respect to the rubric of bioprospecting in the total economic value of biodiversity, David Simpson et al. (1996) of *Resources for the Future* have shown that genetic resources have little value for the pharmaceutical industry (precisely $2.29/hectacre-year in the most biodiverse spot in the world). Although the mathematics of the Simpson model may be impeccable, the end result depends on a scaffolding of assumptions. Just a few years later and in the very same journal, Gordon Rausser and Arthur Small (2000) published a different mathematical model that, lo and behold, shows that genetic resources have very a high value. I count myself among those economists who believe that ours is not a science but a rhetoric; as such any valuation of biodiversity is vaulting ambition which, when plugged into cost-benefit analysis, generates an economics of extinction. For us, the only pertinent question is whether or not there is sufficient *probable cause* to justify public investment in the needed infrastructure to enable a market in genetic resources. Anecdotal evidence such as *Thermus aquaticus,* a microorganism that resulted in a multi-billion dollar industry worldwide, suggests that there is. By enabling a market in biodiversity conservation, stakeholders would be created who could do battle in the political arena where vested interests are hell bent on land use conversion.

IV FIRST, GETTING THE NAMES OF THINGS RIGHT

If economics is rhetoric, then how does the economist proceed to analyze bioprospecting? And how should the journalist report it? There is much to be gained from the wisdom of evolutionists and Dobzhansky is not alone. E.O. Wilson (1998) of Harvard is fond of pointing out that biology begins with getting the names of things right and Richard Dawkins (1995) of Oxford describes genes as pure information. A huge literature exists in the economics of information that could be applied to biodiversity. Economists will argue that the costs for creating information (R&D), are exorbitant while those for reproducing it (manufacturing), negligible. Without a monopoly intellectual property right, everyone would wait for someone else to innovate and few information goods would emerge. So, patents, copyrights, and plant breeder's rights reflect a social contract by which innovators can take a gamble that the fruits of their creativity will have sufficient market value to recover the costs of their R&D.

Examples of getting the names of things right, as Wilson insists, are provided in the boxes below.

Not Quite Synonymous: A BioLexicon of the Debate

Box 2. The Northern Dialect

Biodiscovery: "the collection of samples of biological material; the discovery of bio-active compounds in those samples; and the development of a bio-product, such as pharmaceuticals, based on those bio-active compounds." (www.aar.com.au)

Bioprospecting: "The developing field wherein biologists, chemists, and other researchers are compiling a database of the commercial potential of many species." (www.environment.jbpub.com)

BioTrade Initiative: "Its mission is to stimulate trade and investment in biological resources to further sustainable development in line with the three objectives of the Convention on Biological Diversity…conservation of biological diversity; sustainable use of the components; and fair and equitable sharing of the benefits arising from the utilization of genetic resources" (www.biotrade.org)

Box 3. The Southern Dialect

Biofraud: "The contracting of biodiversity and/or traditional knowledge without having paid an agreed economic rent to all who could have supplied the same input." (www.thebiodiversitycartel.com)

Biopiracy: "The appropriation of the knowledge and genetic resources of farming and indigenous communities by individuals or institutions seeking exclusive monopoly control (usually patents or plant breeder's rights) over these resources and knowledge." (www.amazonlink.org)

Box. 4 Neologisms in the light of economics

Biobetrayal: Northern conservationist NGOs promoting Material Transfer Agreements (MTAs); Southern competent national authorities signing off on those MTAs

Biograb: Identification of a vulnerable country or a community, which will grant prior informed consent to access its genetic resources or traditional knowledge.

Biolooting: Unlimited access to the genetic resources of a country or the traditional knowledge of a community through a comprehensive MTA.

Bioridiculous: The typical royalty of 0.5% (one half of one percent) on net revenue

Biospeak: Calling genetic information, access; calling the price of that information, benefits; and calling the sale "access and benefit-sharing"

Inasmuch as biodiversity involves high opportunity costs in its conservation and negligible costs in its collection, a *quid pro quo* implies that natural information receive a protection equal to that enjoyed by artificial information. Because genes are diffused across taxa and species, across habitats, few countries could ever command monopoly power over biodiversity. What is needed is an oligopoly over natural information, which in plain English, is a biodiversity cartel. The cartel would fix the royalty rate at something significant, say, 15% on net sales, and then distribute the royalties among countries of origin proportional to the size of the habitat of the bioprospected species in each country. Such economic thinking is now taking root in the declarations of The Group of Like-Minded Megadiverse Countries. (Figure 2)

Figure 2 Megadiverse countries: The Group of Like-Minded Megadiverse Countries represents 70% of the terrestrial biodiversity of the planet. Of the seventeen most biodiverse countries, Australia, Russia and the U.S. are not members of the Group.

Contrary to the very spirit of The Group, the Secretariat to The Convention on Biological Diversity has been promoting competition among biodiverse countries by streamlining the procedure for "prior informed consent" to genetic resources. In October 2001, the first meetings took place for "The Bonn Guidelines on Access to Genetic Resources and the Fair and Equitable Sharing of Benefits Arising out of their Utilization." By April 2002, a draft of those guidelines was adopted by COPVI, held in The Hague. An economist would think that the magnitude of the royalties would have been at center stage in both the discussions and the final version of the text. But no, The Bonn Guidelines merely suggest that royalties be negotiated on a case-by-case basis. Inasmuch as the typical royalty in current Material Transfer Agreements (MTAs) is a picayune 0.5% (one half of one percent), the drafters chose to make light of the whole issue. Indeed, royalties [letter (d) of Category 1 of Appendix II of The Bonn Guidelines] is given no more prominence than "Access fees/fee per sample collected" [letter (a)]. Even more disturbingly, the list of monetary benefits [letters (a)-(j)] is followed by a much longer list of non-monetary benefits: capacity-building, technology transfer, and the like [letters (a) through (q) of Category 2]. The impression is unmistakable: little money will change hands in bioprospecting and be happy with those non-monetary benefits! However, economists will not be happy campers. By virtue of the benefits of Category 2 being non-monetary, measurement of their value is all but impossible [e.g. (n): "Institutional and professional relationships..."] and the possibilities for fraud seem infinite. Even if one were to generously assume good faith on the part of all parties, such non-monetary benefits would be a form of earmarking and earmarking is anathema in the economics of public finance. Earmarking precludes the allocation of the budget to those activities with the highest social return.

Could the drafters of The Bonn Guidelines really have been so ignorant of economic theory? The cynic will suspect just the opposite; the drafters were economically very savvy and simply expressed *unenlightened* self-interests. Through the guidelines, industries in the North will be able to exploit rent-seeking behavior within the competent national authorities of the South and secure the coveted prior informed consent for genetic resources. In other words, politicos in the South will suggest a pet project in which they have a direct or indirect interest and sign away the genetic patrimony of their country and that of their neighbors (remember most species are not endemic). Undoubtedly, civil society will criticize such corruption and the text of The Bonn Guidelines will evolve into perpetuity, thereby providing job security for international bureaucrats. In the bright light of economics, the best thing that can be said in favor of The Bonn Guidelines is that they are voluntary; hopefully, the Group of Like-Minded Megadiverse Countries will choose to think like an economist and trash them.

V. CONCLUSION

In homage to Theodosius Dobzhansky, I would say that economics cannot make complete sense of the sundry (and sordid) facts of bioprospecting; in the end, evolution will trump economics. The visceral reaction of disgust that accompanies every bioprospecting proposal cannot be understood with anything from the economist's toolbox. In what has quickly become a seminal article, I read with much delight in *Nature* that monkeys will throw a tantrum if they do not receive the same reward for the same task (Brosnan and de Waal, 2003). Would any self-respecting monkey tolerate monopoly intellectual property rights for the rich and fierce competition over genetic resources for the poor? In the light of evolution, one can understand the ire that bioprospecting generates. *Homo sapiens sapiens* are just not equipped to assimilate royalties of one half of one percent, some recycled lab equipment, and a month or two of vocational training in its use. Any steadfast creationist who may reject such reasoning can thank the dear Lord for such animal spirits.

NOTES

[i] This chapter is a translation and updated revision by the author of "Nada en bioprospección tiene sentido excepto a la luz de la economía". Revista Iberoamericana de Economía Ecológica. No. 1, October 2004

REFERENCES

Brosnan, S. F., & de Waal, F. B. M. (2003). Monkeys reject unequal pay. *Nature*, *425*, 297–299.

Dawkins, R. (1995). *River out of Eden*. New York: Basic Books.

Dobzhansky, T. (1973). Nothing in biology makes sense except in the light of evolution. *The American Biology Teacher*, *35*, 125–129.

Munasinghe, M. (1992). Biodiversity protection policy: Environmental valuation and distribution issues. *Ambio*, *21*(3), 229.

Rausser, G. C., & Arthur, S. (2000). Valuing research leads: Bioprospecting and the conservation of genetic resources. *Journal of Political Economy*, *108*(1), 173–206.

Revkin, A. (2002, May 7). Biologists sought a treaty; Now they fault it. *The New York Times*, p. D1.

Revkin, A. (1990). *The burning season*. Boston: Houghton Mifflin Company.

Simpson, R. D., Sedjo, R. A., & Reid, J. W. (1996). Valuing biodiversity for use in pharmaceutical research. *Journal of Political Economy*, *104*, 163–185.

Vogel, J. (Ed.). (2000). *The biodiversity cartel*. Quito: CARE. www.thebiodiversitycartel.com

Vogel, J. (1997). The successful use of economic instruments to foster sustainable use of biodiversity: Six case studies from Latin America and the Caribbean. White Paper commissioned by the biodiversity support program on behalf of the Inter-American commission on biodiversity and sustainable development in preparation for the summit of the Americas on sustainable development, Santa Cruz de la Sierra, Bolivia. *Biopolicy Journal*, *2*(PY97005), British Library ISSN# 1363-2450, http://www.bdt.org/bioline/py

Vogel, J. (1994). *Genes for sale*. New York: Oxford University Press.

Wilson, E. O., & Frances, M. P. (Eds.). (1988). *Biodiversity*. Washington, DC: National Academy Press.

Wilson, E. O. (1998). *Consilience*. New York: Alfred A. Knopf.

Joseph Henry Vogel
Professor of Economics
University of Puerto Rico

SECTION III SITES OF CONTESTATION

CATRIONA BONFIGLIOLI

6. MAKERS, MESSAGES AND MOVERS

Genetic Technology, the Media and Society

INTRODUCTION

Since Watson and Crick's report of the discovery of the double helical structure of DNA it has become possible to manipulate human, animal and plant genomes (Watson and Crick, 1953; Glover, 1984; Fukuyama, 2002). The resulting DNA revolution is reshaping medicine, farming and food (Bodmer and McKie, 1994; Day and Humphries, 1997; Omenn, 2000). While many benefits are already apparent and more have been promised, biotechnologies and genetic technologies have also been identified as a source of risk and a threat to our human identity (Rifkin, 1983; Wilkie, 1993; Appleyard, 2000; Fukuyama, 2002).

One particular threat which has been identified is that of genetic determinism, which arises from biological determinism, and is based on the assumption that genes alone determine human outcomes such as disease, intelligence, personality and criminal tendencies (Rose, Lewontin et al., 1984). In 1992, Lippman identified scientific discourses as important reinforcers of genetic determinism. Lippman argued that, by routinely using map and blueprint metaphors, scientists reinforce traditions of determinism in ways which narrow concepts of health and illness, privatize responsibility for health and legitimize a new arena for social control (Lippman, 1992).

In 1995, Nelkin and Lindee argued that the gene had become a cultural icon and thereby a driver of genetic determinism (Nelkin and Lindee, 1995). Following Lippman's new genetic 'cartography', Nelkin and Lindee surveyed news media, films, comics, advertisements and television to reveal how popular discourse was transmitting and reinforcing genetic determinism. Since the publication of Nelkin and Lindee's seminal work, Dolly the sheep has been cloned, GM foods have been delivered to millions of dinner plates around the globe, and the human genome has been mapped. In this chapter, I will argue that, rather than being recognised and resisted, genetic determinism continues to be embedded in a key arena of public discourse: mass media news texts. New research in Australia, Europe and North America shows how mass media news articles are playing a crucial role in reinforcing genetic determinism in modern western society through repeated positive messages about genetics and through textual devices including metaphors and frames (Goffman, 1974; Lakoff and Johnson, 1980; Entman, 1993; Scheufele, 1999). News media coverage of genetic technologies is frequently positive, facilitating the acceptance of many genetic technologies into modern life and

Naomi Sunderland et al. (eds.), Towards Humane Technologies: Biotechnology, New Media and Ethics, 77–95.

enhancing the normalization of genetic explanations for human problems. However, as the range of practical applications of genetic technology grows so do the exceptions to this pattern of positive coverage. Notable examples include: cloning, genetic engineering, genetically modified (GM) foods, and, in Iceland, DNA databanking. While these exceptions provide important examples of genetic technologies which attract more negative news, media coverage tends to be dominated by positive framings of genetic technologies, particularly medical genetic technologies, arguably acting as a Trojan horse for genetic determinism. I will also argue that these news media framings play an important role in society because of their intersection with the public's health behaviours and sociocultural and medical phenomena including social attitudes to medical screening and testing and unfolding scientific uncertainties in the genetics of human disease.

Genetic technologies are being developed in a social context which has been described as the 'risk society' in which citizens have been primed to fear new technologies by nuclear accidents, pesticide contaminations and food health scares such as the mad cow disease scandal (Beck, 2000; Carson, 1963; Cooke, 1998; Goodfield, 1981). In a 'risk society' citizens expect, and are expected, to engage in the 'technology of the self' by identifying and controlling risks to themselves and their families (Martin, Gutman et al., 1988; Beck, 2000; Ettorre, 2002; Polzer, Mercer et al., 2002; Petersen, 2006; Bates, 2005). How are citizens deciding which genetic technologies to use? Although public knowledge of genetic technologies is increasing (Voss, 2000), and GM foods are in widespread use, personal experience of other genetic technologies is still relatively rare. For example, it was recently estimated that among the 6.57 million residents of New South Wales, Australia, about 1,800 people had genetic tests in 2000 (Mackintosh and Parr, 2004; Robotham, 2004). Citizens with little direct experience of a phenomenon are more susceptible to media framings of that issue (Baum, 2002). Thus citizens are perceiving the new genetic technologies through the twin lenses of the 'risk society' and the news media (Petersen, 2002; Bonfiglioli, 2005).

MEDIA POWER

Although the news media are not a citizen's sole source of information about genetics and biotechnology, they are recognised as a key source of health information (Wade and Schramm, 1969; Johnson, 1998). News media reach millions of people (Nielsen, 2001; Roy Morgan Research, 2003). Up to 50% of people pay attention to medical news (Roper Starch, 1997; Johnson, 1998). News media help to set the public and policy agenda (McCombs and Gilbert, 1986; Gamson, 1992; Greenfield, 2001); although it has been argued that it is scientists, not journalists, who set the science news agenda (Krimsky, 1982; Goodell, 1986). As Beck argues, the media are essential to the process of defining hazards as risks, and putting them onto the social agenda (Beck, 2000). While the media can insert an issue into the political agenda (McCombs and Gilbert, 1986) and shape the public's understandings of the issue (Entman, 1993), they also reflect public opinion (McCombs and Gilbert, 1986; Gamson and Modigliani, 1989) and will be understood in different ways by different audiences (Hall, 1973) depending on factors such as their personal experience (Baum, 2002). News media audiences will not respond in sheep-like ways, because they decode media messages according to their own values, beliefs and knowledge of the world (Hall, 1973).

Thus genetic technologists navigating the increasingly perilous channel from invention to social acceptance must fight for their cause in the news media arena, which is crucial to shaping social attitudes to genetic technologies (Petersen, 2002; Bonfiglioli, 2005). News media coverage provides a key battleground on which stakeholders struggle to achieve dominance for their interpretations (Ryan, 1991; van Dijck, 1998; Petersen, 2001). The stakes are high: billions of dollars have been invested in genetics and biotechnologies. Negative news coverage may affect consumer uptake, share prices, public policy and research funding.

The strength of news media influence is illustrated by evidence of media effects on the health behaviours of segments of news media audiences.

Recent examples include:

After Australian singer Kylie Minogue's breast cancer was disclosed, average daily TV news coverage of breast cancer increased 20-fold and bookings for mammograms for women who had not been screened before doubled (Chapman, McLeod et al., 2005).

News of links between hormone replacement therapy (HRT) and increases in breast cancer and cardiovascular disease was followed by reductions in the use of HRT of between 18% and 58% (Lawton, Rose et al., 2003; Haas, Kaplan et al., 2004).

In 2000, researchers reported that MMR vaccination fell by 13.6% in an area of Wales reached by negative newspaper coverage about the combined vaccine. The reduction in other areas was only 2.4% (Mason and Donnelly, 2000).

In 1997, a Baltimore Sun newspaper report about a gene linked to colon cancer prompted 250 people to seek advice about testing (Holtzman, Bernhardt et al., 2005).

In 1995-1996, widely publicised government warnings about blood clot risks to users of certain oral contraceptives were followed by a 9.9 per cent increase in pregnancy terminations (Child, MacKenzie et al., 1996).

Almost 60 per cent of 2,256 adults polled on behalf of the US National Health Council said they had changed behaviour after seeing a media health story (Johnson, 1998).

A recent Australian study found an association between increased suicide rates and media coverage of suicide (Pirkis, Burgess et al., 2006).

While only one of these examples is related to genetic technology, they demonstrate the power the news media can have over the health behaviour of some sections of their audiences. Acting on risk information is a fundamental part of health behaviour. It is being constructed as particularly important in the prevention of disease (Khoury and The Genetics Working Group, 1996). Some social scientists argue that we have entered a new age of 'genetic responsibility' in which genetic technologies are transforming our understandings of ourselves and our bodies, and are imposing new responsibilities on individuals to know their genetic risk and deploy the risk information to protect themselves and their families from disease (Rose, 2001; Ettorre, 2002; Polzer, Mercer et al., 2002; Bates, 2005; Lock, 2005).

While it is argued that the news media are a crucial part of citizens' making sense of health risks, the news media are not a public information service. While they are considered an essential tool of democracy, most news media run on commercial grounds, presenting information in an enticing way in order to deliver audiences to advertisers (Conley and Lamble, 2006). Journalists are under pressure to provide attractive articles by selecting subjects with strong news angles and shaping them to maximise their ability to 'grab' the audience's attention (Hurst and White, 1994).

The media do sensationalise some issues and sideline others and sometimes they make mistakes. They are also the target of lobby groups, governments, activists, public relations exponents and others hoping to put their particular spin on the next story. Stakeholders with little awareness of news practices may be irritated or disappointed by the way certain stories appear. This often results in criticism of news practice, even if all the facts are correct and no-one has been misquoted (Bell, 1991). Thus the media constitute an important body of public discourse about genetics and biotechnology, the study of which provides valuable insights into the ways these technologies are being socially constructed.

GENETIC TECHNOLOGIES AND THE MEDIA

How have the news mass media represented biotechnology? News media have communicated excitement about the promise of genetic technology, fears about the risks, and the drama of the debate over gene foods, cloning and stem cells and the race to map the genome (Durant, Hansen et al., 1996; van Dijck, 1998; Bonfiglioli, 2005).

What is intriguing is that the media have been criticised both for being too positive and for being too negative about genetic technology and biotechnology. On the one hand, the news media are accused of being too supportive and unquestioning of genetic technology (White, 1998; Moynihan and Sweet, 2000; Petersen, 2001), being reluctant to report risks or negative or inconclusive findings (Petersen, 2001; Bubela and Caulfield, 2004), encouraging excessive faith in genetic testing (Freedman, 1997) and presenting genetic medicine as offering treatments and prevention in the near future (Bernhardt, Geller et al., 2000).

On the other hand, the news media are accused of demonising genetics (Milmo, 2002). In 2001, Professor John Mattick's keynote speech to the Australian Institute of Health Law and Ethics conference in Melbourne, was entitled 'The Demonisation of Genetics – inequality or inégalité'. In a presentation to the Australasian Medical Writers Association's 1998 conference in Melbourne, Professor Bob Williamson said the media were ascribing negative characteristics to genetics.

Other criticisms of coverage of genetics include omitting important information (Bernhardt, Geller et al., 2000; Holtzman, Bernhardt et al., 2005), focusing on extreme scenarios to the detriment of calm debate (Caulfield, 2000), promoting rapid introduction of genetic tests (Kodish, 1997), focusing too much on behavioural genetics (Bubela and Caulfield, 2004) and promoting genetic determinism (Lippman,

1992; Petersen, 2002). The media are often accused of sensationalising issues (Hurst and White, 1994), such as health scares (Harrabin, Coote et al., 2003), but an analysis of newspaper articles about genetics found only 11% of articles were exaggerated and 82% contained no significant technical errors (Bubela and Caulfield, 2004).

Accused of both positive and negative bias, it may be that the media are achieving some degree of balance. Despite accusations of demonisation, analyses of the balance between negative and positive coverage usually find positive coverage dominates (Priest and Talbert 1994; White, 1998; Cormick, 2001; Petersen, 2001; CARMA International, 2002), arguably a sign of public relations activity. One study has found negative coverage dominated (Pálsson and Harðardóttir, 2002). In a recent Australian study of orientation in newspaper coverage of genetic technologies between 1992 and 1999, positively oriented articles outnumbered negative items in every year studied (Bonfiglioli, 2005). However, the proportion of articles which were negative doubled between 1998 to 1999. This was largely explained by the rapid increase in 1999 in the proportion of articles which were about genetically modified foods rather than about genetic medicine, and the intense public debate about the labelling of GM foods which took place in 1999 (Bonfiglioli, 2005). In 1999, 46% of GM food newspaper articles were negatively oriented while 45% of articles about genetic medicine were positive (Bonfiglioli, 2005).

Analyses of news media orientation conducted for Biotechnology Australia also found positive coverage tended to dominate (Cormick, 2001). However, a survey of impressions of GM food coverage in 2000 found that almost 50% of people questioned believed the coverage to be negative even though only 34% of coverage at that time was negative. Only 7% of people said they thought gene food coverage was positive, compared with 18% of actual coverage being positive (Cormick, 2001). This suggests that even when the news media publish positive items, the negative items are the ones that stick in the mind.

It is not clear why news media audiences perceive coverage to be more negative than it is. It could be argued that in today's risk-sensitive society negative coverage is more salient and therefore more memorable than positive news. News editors favour negative news because it sells more newspapers. People may be more likely to remember negative news because neurotransmitters released in response to negative experiences may strengthen memory formation (Pitman, Sanders et al., 2002). For scientists, the problem could be that they have been enjoying years of relatively respectful, largely positive coverage and that any negative coverage is indicative that scientists' status in society has been undermined by technological failures (Goodfield, 1981: pp.3-4). An observation that positive news clusters in the finance pages, where press releases are often used as source material, may help to explain differences between perception and actual coverage as finance pages are not read by all readers (Cormick, 2001).

It is also possible that the community's sensitivity to issues of food safety means that news about genetically modified (GM) food resonates more strongly than other genetic news (Wilkie and Graham, 1998; Gaskell, Bauer et al., 1999; Nuffield Council on Bioethics, 1999). Recent Australian research shows that GM food is the subject of much more negative coverage than genetic medicine, particularly in 1999 during the GM food debate (Bonfiglioli, 2005). Petersen's analysis (Petersen, 2001), which found such a positive presentation of genetic science, did not include articles about GM foods. An in-depth analysis of three Australian newspapers' coverage of GM foods found negative framings dominated the discourse. The case against GM foods was constructed using frames of controversy, regulation, environmental risk and fear for human health (Bonfiglioli, 2005). The most dominant frames supporting GM foods were reassurance, wonders of science, and denigration of calls for GM foods to be labelled (Bonfiglioli, 2005).

GM foods are certainly not the only subject of negative coverage about genetic and biotechnologies in the news media. Cloning has been a focus for negative news at least since Rorvik (1978) falsely claimed the first human clone had been born, causing a public furor (van Dijck, 1998) as did the splitting of human embryos (Kolata, 1993). Dolly the cloned sheep generated an explosion of news media coverage (Bonfiglioli, 2005). The controversial breakthrough in animal technology was largely discussed in terms of human cloning (Turner, 1997; Hodgson, 1998; Hotz, 1998). A recent analysis of three Australian newspapers' coverage of Dolly's cloning showed that of 75 news articles published in the months after the announcement 52 (69%) linked animal and human cloning (Bonfiglioli, 2005).

Despite the emphasis on human cloning, a surprisingly small proportion of articles were predominantly negative: falling from 25% in 1997 to less than 10% in 1998 and 1999. This suggests that while cloning was a cause of concern, some newspaper articles provided venues in which scientists could draw boundaries between 'good' therapeutic cloning (particularly that used to develop stem cells) and 'bad' cloning used to make copies of human beings. These distinctions reaffirmed the positive image of genetics and left the impression that cloning research can have significant medical benefits (Petersen, 2001; Bonfiglioli, 2005).

Although both GM foods and cloning are the focus for negative sentiment towards genetic technologies, these issues are embedded in a media environment of good news stories about genetic medicine. The contrast between GM food coverage and genetic medicine coverage is consistent with European observations that people distinguish between 'red' biomedical genetic technologies which they strongly support and 'green' agribiotechnologies about which they have reservations (Bauer, 2002). In the next section I will examine a key aspect of news media discourse about medical genetic technologies – genetic determinism.

GENETIC DETERMINISM

Although it can be argued that virtually every disease has a genetic element (Collins, 1999; Khoury, Burke et al., 2000), a person's physical and social environment also plays a vital role both in initial physical development and in the expression of disease (Lewontin, 1992; Hubbard and Wald, 1999; Greenfield, 2003). However, while the importance of environmental factors is widely recognised by scientists, some scientists, doctors, patients, families, advocacy groups and members of the public continue to be 'captivated by genetic determinism' and genetic explanations tend to be prioritised in public and scientific discourse; thus diverting attention from non-genetic factors (Willis, 2002; Lock, 2005). This kind of geneticisation is reinforced by the widespread use of deterministic language (Lippman, 1992; Nelkin and Lindee, 1995; Hubbard and Wald, 1999; Nerlich, Dingwall et al., 2002). Media coverage which foregrounds genetic influences and downplays environmental factors reflects, generates, and reinforces genetic determinism (Nelkin and Lindee, 1995; Conrad and Weinberg, 1996; Conrad, 1997; Conrad and Gabe, 1999; Nelkin and Lindee, 1999; Conrad, 2001; Petersen, 2001).

News coverage may contribute to determinism by often referring to 'the' gene for a disease or a behaviour, by confusing genetic markers with genes, by employing the frame of 'genetic optimism', and by portraying genetic discoveries as likely to revolutionise medicine (Alper and Beckwith, 1993; Conrad, 1999; Conrad, 2001; Conrad and Markens, 2001). While a single gene can determine, disease in rare, monogenic disorders such as Huntington's, in most diseases genes contribute to, but do not determine illness. When genetic markers are discovered for traits such as homosexuality, the news media wrongly refer to the marker as 'the gay gene' (Conrad and Markens, 2001). The media's use of 'genetic optimism' framing, which constructs genetic discoveries as likely to lead to effective treatments, also contributes to determinism (Conrad, 2001; Conrad and Markens, 2001).

A major longitudinal study of the prevalence of genetic determinism in public discourse including media coverage (Condit, Ofulue et al., 1998; Condit, 1999) has found that one third of coverage attributed disease or behaviours to genetic causes alone. However, the study found the use of deterministic discourse fell in mass magazines and remained relatively static in other media coverage (Condit, Ofulue et al., 1998). However, another study made the crucial observation that, while genetic discoveries receive prominent news coverage, the media tend to neglect reports of disconfirming evidence, thus creating cultural residues in people's minds so that they remember the original genetic link to a disease or a behaviour but are not aware of the contradictory evidence (Conrad, 2001).

In Australia, newspaper coverage of genomics in the year 2000 was also found to make widespread use of deterministic language (Bonfiglioli, 2005). In a study of three major newspapers' coverage of the announcement of the mapping of the Human Genome in 2000, more than 90% of articles employed deterministic language, such as map, code or book metaphors (Bonfiglioli, 2005). These metaphors create a sense of the genome as readable and re-writable: once you crack the code you can use it to reshape life (Lippman, 1992; Nerlich, Dingwall et al., 2002). The news articles in this sample were three times more likely to refer to genes causing disease, behaviour or other phenomena than to refer to the pivotal role the environment plays in the development of most human traits. Most articles did not refer to the risks of genetic determinism (Bonfiglioli, 2005).

News coverage appears to be an important arena for deterministic discourse which may inadvertently exaggerate the benefits of genetic technologies, inflate expectations about the promise of genetic medicine and contribute to a lack of appreciation of the importance of social and physical environments in the development of disease and behavioural traits. This may be attributed in part to the nature of journalism which judges issues on their ability to meet news values criteria such as extraordinariness, impact, prominence, and timeliness, and favours episodic rather thematic news: "Genes are far more newsworthy than social or economic circumstances as a source of antisocial behaviour" (Iyengar, 1991; Nelkin and Lindee, 1999: p.159). Although news coverage carries some arguments against genetic technologies, debates are framed in genetic terms thus contributing to the perception that genetic technologies will inevitably be used.

Condit's analysis suggests the use of genetically deterministic language is falling in response to increased recognition of the complexities of genetic effects, environmental influences and social factors (Condit, 1999). However, the discourse of genetic determinism continues to be a salient, perhaps dominant, feature of the discourse landscape and as such has important implications for future health behaviours and policy making (Bates, 2005).

The genetic paradigm promotes genetic solutions, obscures the potential risks of genetic technologies and biotechnologies and undermines non-genetic approaches to disease and social problems including public health, environmental and social reforms (Lippman, 1992; Nelkin and Tancredi, 1994). Highlighting the power of the gene while downplaying environment and lifestyle may exacerbate current trends of preferring medication to lifestyle modification; with important implications for health budgets (Welsh Health Planning Forum, 1995; Callahan, 1998).

Social scientists argue that genetic determinism, also known as geneticisation or 'genetic essentialism', underpins new trends in political conservatism (Rose, Lewontin et al., 1984; Krimsky, 1991; Nelkin and Tancredi, 1994; Nelkin and Lindee, 1995). Genetic determinism promotes victim blaming, privatization of disease prevention and reductionism, as well as Cartesian concepts of the body and mind as machines (Lippman, 1992; Nelkin and Tancredi, 1994; Conrad, 1999). Genetic essentialism appeals as a justification for passive attitudes towards social injustice and implies that there are genetic solutions to social dilemmas; "suggesting that the improvement of society depends, ultimately, on the improvement of DNA" (Lippman, 1992; Nelkin and Lindee, 1999: p.165).

THE SIGNIFICANCE OF NEWS COVERAGE TRENDS

The significance of the framing of genetic technologies in the media discourse is reinforced by considering the intersection between genetic determinism in media discourses and key aspects of genetic technologies. Citizens whose understanding of genetics is likely to be largely shaped by media discourse are increasingly being expected to make choices about using genetic technologies. Genetic technologies are being marketed to the public, sometimes directly, at a time when the field of epigenetics is producing new understandings of the complexities of gene expression, the social and commercial forces which drive uptake of genetic technologies are spreading and the uncertainties surrounding the interpretation of genetic test results are being increasingly recognised. In the following section, I will outline key aspects of epigenetics, trends in testing, consumer and commercial influences, genetic test uncertainties and touch on the issue of genetic discrimination.

Epigenetics and Gene Penetrance

The Human Genome Project focused on identifying genes on the assumption that genes control human biology. As well as finding a surprisingly small number of genes (about 30,000 instead of the 100,000 predicted), genome mapping identified large swaths of DNA which were labelled as 'junk' because they did not contain what were recognised as genes. New research is challenging these assumptions by revealing that junk DNA codes for RNA which acts directly on DNA, other RNA and proteins. These segments of RNA appear to play a crucial role in gene regulation, gene expression, and gene silencing, as well as allowing different cells to produce different proteins by controlling how RNA sequences are spliced together to form different proteins from the same sequence of DNA (Mattick, 2004).

These discoveries may help to explain why some people with disease susceptibility genes do not develop the disease (Lock, 2005). They have also revived 19th century theories developed by the French scientist Jean-Baptiste Lamarck who posited that acquired characteristics could be inherited (Palmer, 2005). This concept was widely discredited by the development of classical genetics and Darwinian evolutionary theory. The discovery of the structure of DNA in 1953 was thought to have sealed the fate of Lamarckianism by establishing the central dogma that genetic information flows in one direction from the DNA to the RNA to the protein (Mattick, 2004; Lock, 2005). However, in the 1980s, Marcus Pembrey found that some human genes are 'imprinted' with an epigenetic marker which showed which parent that gene comes from; in 1998 it was argued that evidence from the immune system supported the existence of a Lamarckian mechanism (Steele, Lindley et al., 1998); and, more recent discoveries in epigenetics suggest that environmental exposures, including conditions experienced in the womb, can leave a genetic mark on the genome which can be inherited (Mattick, 2003; Lock, 2005). Epigenetic research also suggests that non-coding RNA is playing an important role in the differential expression of genes (Mattick, 2003). Epigenetic discoveries are challenging widely accepted understandings of the relationship between susceptibility genes and disease and undermining efforts to estimate probability of individual risk (Lock, 2005).

Trends in Testing

While many genetic tests are offering benefits to newborn babies and people with a family history of genetic disease, there are trends in testing, both genetic and non-genetic, which could be a cause for concern as more and more genetic tests are marketed, sometimes directly to the public. The dominant discourse of individual choice and the concomitant personal responsibility this discourse infers are driving forces behind people seeking to obtain and act upon risk information, including through genetic and other health tests (Polzer, Mercer et al., 2002). Individuals are engaging in this 'technology of the self' to reduce their risk of genetic and other diseases and to fulfil their obligations to protect their families from inherited disease (Martin, Gutman et al., 1988; Ettorre, 2002; Polzer, Mercer et al., 2002;

Petersen, 2006). In the developed world, there are powerful social and commercial forces placing pressures on citizens to embrace genetic and health tests. These pressures include increasingly defensive medicine, normalizing of prenatal genetic testing, biotechnology companies' need to recoup their investments, and the marketing of tests to people through the internet and by mail (Petersen, 1998; Ettorre, 2002; Williams-Jones, 2003). A notable example of new pressure to test is the US surgeon general's initiative to encourage people to collect family health data and share it with their doctors. This family health tree project may undermine public health advocacy and promote genetic determinism because it downplays the importance of environmental and behavioural factors in the development of common diseases which have a genetic component (Bates, 2005).

A powerful example of consumer demand for health tests was the fivefold increase in prostate specific antigen (PSA) testing between 1989 and 1996 (Smith and Armstrong, 1998). In that time, more than 1.1 million Australian men had a PSA test, adding up to a large uncontrolled experiment in screening. Although mortality from prostate cancer is falling, it is still debated whether the benefits of PSA screening outweigh the risks (Albertson, 2003). The PSA testing phenomenon demonstrates that even when experts recommend against screening healthy people for disease risk, commercial and social pressures may lead to high test uptake whether or not there is evidence of benefit. There are also suggestions that although breast screening reduces mortality, women's decisions to undergo mammography may be being swayed by information leaflets given to women in New South Wales, Australia, which have been criticised for over-emphasising the benefits, failing to explain potential risks adequately and conveying the idea that participation was a public duty (Robotham, 2006).

New research shows that some people are taking genetic tests under pressure from their families, insurance companies and mortgage lenders (Robotham, 2004) and many are doing so in free market economies where genetic tests can be marketed directly to the public even in the absence of benefit (Lock, 2005). The psychosocial impacts of some genetic tests may be considerable, including genetic discrimination, increased risk of suicide, and restricted access to insurance and employment (Billings, Kohn et al., 1992; Almqvist, Bloch et al., 1999).

Commercial and Research Influences

Non-genetic medical tests are already widely marketed and genetic test makers have begun to use direct mail and the internet to offer testing to the public, with private testing for susceptibility to genetic disease predicted to rise (Gollust, Wilfond et al., 2003; Williams-Jones, 2003; Lock, 2005). Tests offered include haemochromatosis, cystic fibrosis, as well as nutrition, behaviour, Alzheimer's disease susceptibility and aging tests. Of 14 websites offering health-related genetic tests, less than half mentioned risks associated with genetic services or the availability of counselling (Gollust, Wilfond et al., 2003). Direct-to-consumer test services are of concern because people may decide to get tested without considering what the results will mean in terms of disease or other impacts. Other genetic tests such as paternity testing and identity testing are readily available through more than 100 websites with their potential for harm recently being recognised (Gollust, Wilfond et al., 2003).

Researchers in Australia and the US are trialling workplace screening for hereditary haemochromatosis, a disorder of iron storage which leads to organ damage and early death. Although the researchers report no ill effects (Nisselle, Delatycki et al., 2004; Adams, 2005) there are debates about what proportion of people who have two copies of the main genetic variation linked to haemochromatosis will develop disease (Beutler, Felitti et al., 2002; Adams, 2005; Waalen, Nordestgaard et al., 2005) and researchers are investigating the role of other genetic and environmental factors, particularly alcohol, which may influence whether disease develops (Waalen, Nordestgaard et al., 2005). Workplace settings may undermine the voluntary nature of testing and fail to protect privacy (Schmitz and Wiesing, 2006).

In addition, researchers are conducting trials of tests for susceptibility to conditions such as Alzheimer's disease (Lock, 2005). Although these trials may relieve anxiety in participants with a family history who are found not to have the relevant genetic variant, there are no immediate benefits in terms of prevention or treatment (Lock, 2005). Such studies may provide valuable insights into how people deal with knowing their genetic status but uncertainty about how many people with the disease-linked variant actually develop the disease means some people will worry more but never develop that condition (Lock, 2005).

Genetic Test Result Uncertainties

Another key concern is that there are uncertainties in the information on which clinicians base risk assessments for genetic diseases such as familial breast cancer, haemochromatosis and Alzheimer's (Lock, 2005). Although the well-publicised breast cancer genes BRCA 1 and 2 are linked to very high rates of disease, it has been argued that most familial breast cancer is likely to be accounted for by genes with less power to cause disease (Nathanson, Wooster et al., 2001). In haemochromatosis, more research is needed to resolve debates about disease penetrance (Beutler, Felitti et al., 2002; Adams, 2005; Waalen, Nordestgaard et al., 2005). In early onset familial Alzheimer's disease, the age of onset can vary in twins by as much as 20 years, which suggests that important factors other than genes must also be involved in disease expression (Lock, 2005). In late-onset Alzheimer's disease people with two copies of a particular version of the APOE gene are at greatly increased risk but many people with the 'risky' genetic profile do not get Alzheimer's and many people who do get the disease do not have this genetic variant, again suggesting other genes and the environment are influencing disease expression (Lock, 2005). Lock criticises the use of such uncertain information for adult onset disease as being 'no more accurate than fortune-telling' and for having no personal utility (Lock, 2005: p.S48).

Differing estimates of rates of penetrance – the proportion of people with a particular genetic variant who will actually develop the disease linked to that variant – may arise because original estimates were based on families with high penetrance genetic variants, because epigenetic factors influence disease expression, or because environmental factors and other genes influence disease development (Nathanson, Wooster et al., 2001; Willis, 2002; Lock, 2005). The concern is that if citizens gain their understandings of the powers of particular genes from media coverage, which by its nature highlights powerful genes, how will they interpret test results which show they have a gene of concern? Questions also arise about health care service protocols for identifying who is at high risk and who should be invited to have genetic tests (Prior, 2001). Many thousands of American women have already had their breasts and ovaries removed to avoid what they may see as a certain fate. These preventive actions have increased lifespan (Griffith, Edwards et al., 2004) but the question remains whether some of these women might have instead chosen a high-surveillance routine if they had believed their risk to be lower than the high rates so widely publicised. The role of the media in decision making requires further exploration after a study found that 75% of newspaper articles about BRCA 1 referred to prophylactic mastectomy but only 15% referred to treatment of breast cancer (Bernhardt, Geller et al., 2000).

Genetic Discrimination

The potential psychosocial risks of genetic testing have been well explored in the research literature (Marteau and Richards, 1996; Almqvist, Bloch et al., 1999; Chadwick, Shickle et al., 1999) and should continue to be monitored and more widely reported in the media. New evidence of genetic discrimination in insurance and employment continues to emerge (Billings, Kohn et al., 1992; Low, King et al., 1998; Barlow-Stewart and Keays, 2001). The emphasis insurers place on biomarkers such as high blood cholesterol does not bode well for citizens if insurers are given free rein to ask for genetic tests (Van Hoyweghen, Horstman et al., 2006). In Australia, the Australian Law Reform Commission and the Australian Health Ethics Committee have investigated genetic testing in great depth but citizens are waiting for these findings to be translated into the stronger protection they may require as genetic testing becomes more common (Australian Law Reform Commission, 2003).

CONCLUSION

This chapter has outlined the importance of the news media to citizens' understandings of genetic technologies and biotechnologies, described several key features of media coverage of genetic and biotechnologies, and touched on the cultural landscape into which genetic technologies are being introduced. While media coverage is by no means the only arena in which the discourses of genetic determinism and optimism are located, news media remain an influential source of health information.

On the one hand, powerful media, public and commercial discourses are framing genetic technologies as potent new products to control disease and reform behaviour. On the other hand, diseases are being influenced by a complex combination of genetic, environmental and lifestyle factors. In the middle, citizens are increasingly expected to exercise individual responsibility and 'choice' in identifying and addressing genetic risks to their health.

How can the situation be improved? Citizens, health care providers, policymakers, insurers, and employers need continued education about the nature of genetic technologies including the risks. This can be achieved to a limited extent through enhanced interaction with the news media but it is important to recognise that the institutional practices of the media mean that genetic news, like other health news, is handled in a way which may be too simple, too dramatic, or too polarised to be useful as the sole source of information for the public. While the internet can be a source of excellent information, such good information is surrounded by a sea of less-reliable content. As with the news media, websites are also being colonised by public relations generated material.

Teachers, scientists and health organisations are already working hard to develop information resources for citizens, but citizens need signposts to point them to reliable sources which can provide them with high quality information on which they can make sound decisions. Not everyone has access to the internet, so efforts to include genetic education within school curricula should be extended to ensure all citizens leave school with some understanding of the benefits, risks and complexities of genetic technologies. People developing information products such as leaflets should be aware of the way presentation of information can influence citizens' attitudes towards genetic testing – one study has found using a 'glossy' format made people feel more positive about genetic testing (Michie, di Lorenzo et al., 2004).

Health professionals and policymakers need to be aware that public discourse about health and medicine is being colonised by genetic 'frames', diverting attention and perhaps resources away from non-genetic interventions such as clean water, housing, education and vaccination which have brought massive and ongoing improvements in overall health (Clarke, 1995; Willis, 1998; Willis, 2002). Media coverage of genetic technologies is replete with genetic determinism and genetic optimism and at the same time it tends to neglect potential harms of genetic technology and non-genetic factors. It rarely questions the goals, directions, methods, or value of genetic research (Petersen, 2001).

Just as many health screening tests are now being evaluated for safety, cost and benefit, it is advisable that genetic tests should also be evaluated before being funded by the public purse or released directly to the public. While people may need to be alerted to the possible harms of accessing genetic testing from the internet, it should be acknowledged that anonymous internet testing may allow people to feel the testing process is more private. Genetic discrimination may be detrimental to health as well as social justice and needs to be curtailed through public education and legislation to protect people who take genetic tests from being unfairly disadvantaged in insurance, health care or employment (Alper and Beckwith, 1993).

Journalists could benefit from paying more attention to the commercial interests of their sources, ensuring they seek balancing or confirming information and trying where possible to acknowledge non-genetic factors in disease causation, while lobbying editors for greater space in which to explain the issues (Holtzman,

Bernhardt et al., 2005). Media organisations should expand their support for specialist medical and science writers, and try to improve the scientific literacy of all journalists and editors (Holtzman, Bernhardt et al., 2005).

Individuals and organisations who wish to see changes to the ways media cover genetic and biotechnologies can learn more about media practice, help to educate journalists about their field of expertise, be more available for comment and explanation, refrain from exaggeration (Holtzman, Bernhardt et al., 2005) and, most of all, actively engage with news media processes by contacting journalists, writing letters to the editor, and participating in public meetings and online discussions.

In the face of scientific uncertainty and a growing appreciation of the risks alongside the benefits of genetic testing, people are increasingly in need of quality assessment of the risks and benefits of tests, good quality information through media and other sources, and practical and ethical controls over the use of genetic information. If the powerful social discourses of choice and personal responsibility are driving people to embrace genetic technologies, particularly genetic tests, it is incumbent upon society to develop a safe social and regulatory environment. This is crucial so that people who seek genetic testing to protect their own health, or that of their families, are protected from the recognised potential and actual harms of testing: discrimination, stigmatisation, and determinism.

REFERENCES

Adams, P. C. (2005). Screening for haemochromatosis - Producing or preventing illness? *Lancet, 366*, 269–271.

Albertson, P. C. (2003). The prostate cancer conundrum. *Journal of the National Cancer Institute, 95*(13), 930–931.

Almqvist, W. W., & Bloch, M., et al. (1999). A worldwide assessment of the frequency of suicide, suicide attempts, or psychiatric hospitalization after predictive testing for Huntington disease. *American Journal of Human Genetics, 64*, 1293–1304.

Alper, J. S., & Beckwith, J. (1993). Genetic fatalism and social policy: The implications of behavior genetics research. *Yale Journal of Biology and Medicine, 66*, 511–524.

Appleyard, B. (2000). *Brave new worlds: Genetics and the human experience*. London: Harper Collins.

Australian Law Reform Commission. (2003). *Essentially yours: The protection of human genetic information in Australia*. Sydney, NSW: Australian Law Reform Commission, National Health & Medical Research Council, Australian Health Ethics Committee.

Barlow-Stewart, K., & Keays, D. A. (2001). Genetic discrimination in Australia. *Journal of Law and Medicine, 8*, 250–262.

Bates, B. R. (2005). Care of the self and patient participation in genetic discourse: A Foucauldian reading of the surgeon general's "My family health portrait" program. *Journal of Genetic Counseling, 14*, 423–434.

Bauer, M. W. (2002). Controversial medical and agri-food biotechnology: A cultivation analysis. *Public Understanding of Science, 11*(April), 93–111.

Baum, M. (2002). Sex lies and war: How soft news brings foreign policy to the inattentive public. *American Political Science Review, 96*(1), 91–109.

Beck, U. (2000). Risk society revisited: Theory, politics and research programmes. In B. Adam, U. Beck, & J. van Loon (Eds.), *The risk society and beyond* (pp. 211–229). London: Sage Publications.

Bell, A. (1991). *The language of news media*. Oxford; Cambridge, MA: Blackwell.

Bernhardt, B., & Geller, G., et al. (2000). Analysis of media reports on the discovery of breast and prostate cancer susceptibility genes [abstract 268]. *American Journal of Human Genetics, 67*(Suppl. 2), 62.

Beutler, E., & Felitti, V., et al. (2002). Penetrance of 845G--> A (C282Y) HFE hereditary haemochromatosis mutation in the USA. *Lancet, 359*(9302), 211–218.

Billings, P., & Kohn, M., et al. (1992). Discrimination as a consequence of genetic testing. *American Journal of Human Genetics, 50*(3), 476–482.

Bodmer, W. F., & McKie, R. (1994). *The book of man: The quest to discover our genetic heritage.* London: Little Brown.

Bonfiglioli, C. M. F. (2005). *Genetic technology in the Australian media.* PhD thesis, Faculty of Medicine, University of Sydney, Sydney.

Bubela, T. M., & Caulfield, T. A. (2004). Do the print media 'hype' genetic research? A comparison of newspaper stories and peer-reviewed research papers. *Canadian Medical Association Journal, 170*(9), 1399–1407.

Callahan, D. (1998). *False hopes: Why America's quest for perfect health is a recipe for failure.* New York: Simon & Schuster.

CARMA International. (2002). *Media analysis April 2002–May 2002.* Canberra: Biotechnology Australia.

Carson, R. (1963). *Silent spring.* London: Hamish Hamilton.

Caulfield, T. A. (2000). Underwhelmed: Hyperbole, regulatory policy and the genetic revolution. *McGill Law Journal, 45,* 437–460.

Chadwick, R., & Shickle, D., et al. (1999). *The ethics of genetic screening.* Dordrecht, the Netherlands: Kluwer Academic Publishers.

Chapman, S. F., & McLeod, K., et al. (2005). Impact of news of celebrity illness on breast cancer screening: Kylie Minogue's breast cancer diagnosis. *Medical Journal of Australia, 183*(5), 247–250.

Child, T., & MacKenzie, I., et al. (1996). Terminations of pregnancy, not unplanned deliveries, increased as result of pill scare. *British Medical Journal, 313*(7063), 1005.

Clarke, A. J. (1995). Population screening for genetic susceptibility to disease. *British Medical Journal, 311*(6996), 35–38.

Collins, F. S. (1999). Medical and societal consequences of the human genome project. *New England Journal of Medicine, 341*(1), 28–37.

Condit, C. M. (1999). *The meanings of the gene: Public debates about human heredity.* Madison, WI: University of Wisconsin Press.

Condit, C. M., & Ofulue, N., et al. (1998). Determinism and mass-media portrayals of genetics. *American Journal of Human Genetics, 62,* 979–984.

Conley, D., & Lamble, S. (2006). *The daily miracle.* Melbourne: Oxford University Press.

Conrad, P. (1997). Public eyes and private genes: Historical frames, news constructions, and social problems. *Social Problems, 44*(2), 139–154.

Conrad, P. (1999). A mirage of genes. *Sociology of Health and Illness, 21*(2), 228–241.

Conrad, P. (2001). Genetic optimism: Framing genes and mental illness in the news. *Culture, Medicine & Psychiatry, 25*(2), 225–247.

Conrad, P. (2001). Media images, genetics, and culture: Potential impacts of reporting scientific findings on bioethics. In B. Hoffmaster (Ed.), *Bioethics in social context* (pp. 90–111). Philadelphia: Temple University Press.

Conrad, P., & Gabe, J. (Eds.). (1999). *Sociological perspectives on the new genetics.* Oxford, Malden, MA: Blackwell Publishers.

Conrad, P., & Markens, S. (2001). Constructing the 'gay gene' in the news: Optimism and skepticism in the American and British press. *Health, 5*(3), 373–400.

Conrad, P., & Weinberg, D. (1996). Has the gene for alcoholism been discovered three times since 1980? A news media analysis. *Perspectives on Social Problems, 8,* 3–24.

Cooke, J. (1998). *Cannibals, cows & the CJD catastrophe.* Milsons Point, NSW: Random House Australia.

Cormick, C. (2001). *Genetic manipulation or information manipulation.* Canberra, ACT: Biotechnology Australia. 2002.

Cormick, C. (2001). *Scientists need to get out of their laboratories and engage in public debate.* Canberra, ACT: Biotechnology Australia. 2002.

Day, I. N. M., & Humphries, S. E. (1997). *Genetics of common diseases: Future therapeutic and diagnostic possibilities.* Oxford: Bios Scientific Publishers.

Durant, J., & Hansen, A., et al. (1996). Public understanding of the new genetics. In T. Marteau & M. Richards (Eds.), *The troubled helix: Social and psychological implications of the new human genetics* (pp. 235–248). Cambridge, UK: Cambridge University Press.

Entman, R. (1993). Framing: Towards clarification of a fractured paradigm. *Journal of Communication, 43*(4), 51–58.

Ettorre, E. (2002). A critical look at the new genetics: Conceptualizing the links between reproduction, gender and bodies. *Critical Public Health, 12*(3), 237–250.

Freedman, T. (1997). Genetic susceptibility testing: A therapeutic illusion? *Cancer, 79*, 2063.

Fukuyama, F. (2002). *Our posthuman future: Consequences of the biotechnology revolution.* New York: Farrar, Straus & Giroux.

Gamson, W. A. (1992). *Talking politics.* New York: Cambridge University Press.

Gamson, W. A., & Modigliani, A. (1989). Media discourse and public opinion on nuclear power: A constructionist approach. *American Journal of Sociology, 95*, 1–37.

Gaskell, G., & Bauer, M. W., et al. (1999). Worlds apart? The reception of genetically modified foods in Europe and the U.S. *Science, 285*(5426), 384–387.

Glover, J. (1984). *What sort of people should there be? Genetic engineering, brain control and their impact on our future world.* Harmondsworth, Middlesex: Penguin.

Goffman, E. (1974). *Frame analysis: An essay on the organization of experience.* Cambridge, MA: Harvard University Press.

Gollust, S., & Wilfond, B. S., et al. (2003). Direct-to-consumer sales of genetic services on the internet. *Genetic Medicine, 5*(4), 332–337.

Goodell, R. (1986). How to kill a controversy: The case of recombinant DNA. In S. Friedman, S. Dunwoody, & C. Rogers (Eds.), *Scientists and journalists: Reporting science as news* (pp. 170–181). New York: Macmillan.

Goodfield, J. (1981). *Reflections on science and the media.* Washington, DC: American Association for the Advancement of Science.

Greenfield, S. A. (2001). *The brain of the future.* Australian Broadcasting Corporation.

Greenfield, S. A. (2003). Brain function. *Archives of Disease in Childhood, 88*, 954–955.

Griffith, G., & Edwards, R., et al. (2004). Estimating the survival benefits gained from providing national cancer genetic services to women with a family history of breast cancer. *British Journal of Cancer, 90*(10), 1912–1919.

Haas, J., & Kaplan, C., et al. (2004). Changes in the use of postmenopausal hormone therapy after the publication of clinical trial results. *Annals of Internal Medicine, 140*(3), 184–188.

Hall, S. (1973). *Encoding and decoding in the television discourse* (Media Series SP No.7). Council of Europe Colloquy on "Training in the Critical Reading of Television Language". Centre For Contemporary Cultural Studies, University of Birmingham.

Harrabin, R., & Coote, A., et al. (2003). *Health in the news: Risk, reporting and media influence.* London: King's Fund.

Hodgson, A. (1998). *Undressing dolly: A clone's 12 months gestation period in the UK Press.* London and Maastricht: University of East London and University of Maastricht.

Holtzman, N. A., & Bernhardt, B. A., et al. (2005). The quality of media reports on discoveries related to human genetic diseases. *Community Genetics, 8*(3), 133–144.

Hotz, R. L. (1998). Dolly, Dr. Seed, and five issues that need study. *ScienceWriters, Winter*(4), 1–4.

Hubbard, R., & Wald, E. (1999). *Exploding the gene myth: How genetic information is produced and manipulated by scientists, physicians, employers, insurance companies, educators, and law enforcers.* Boston: Beacon Press.

Hurst, J., & White, S. A. (1994). *Ethics and the Australian news media.* South Melbourne: Macmillan Education Australia.

Iyengar, S. (1991). *Is anyone responsible? How television frames political issues.* Chicago: University of Chicago Press.

Johnson, T. (1998). Shattuck lecture - Medicine and the media. *New England Journal of Medicine, 339*(2), 87–92.

Khoury, M. J., & Burke, W., et al. (Eds.). (2000). *Genetics and public health in the 21st century: Using genetic information to improve health and prevent disease. Oxford monographs on medical genetics.* Oxford: Oxford University Press.

Khoury, M. J., & The Genetics Working Group. (1996). Public health policy forum: From genes to public health: The application of genetic technology in disease prevention. *American Journal of Public Health, 86*(12), 1717–1722.

Kodish, E. (1997). Commentary: Risks and benefits, testing and screening, cancer, genes and dollars. *Journal of Law, Medicine & Ethics, 25*(4), 252–255.

Kolata, G. (1993). Human embryos cloned, US experiment produces identical twins and triplets. In *The Sydney Morning Herald* (p. 1). Sydney, NSW.

Krimsky, S. (1982). *Genetic alchemy: The social history of the recombinant DNA controversy.* Cambridge, MA: MIT Press.

Krimsky, S. (1991). *Biotechnics & society: The rise of industrial genetics.* New York: Praeger.

Lakoff, G., & Johnson, M. (1980). *Metaphors we live by.* Chicago, Illinois: The University of Chicago Press.

Lawton, B., & Rose, S., et al. (2003). Changes in use of hormone replacement therapy after the report from the women's health initiative: Cross sectional survey of users. *British Medical Journal, 327,* 845–846.

Lewontin, R. C. (1992). *Biology as ideology.* New York: Harpers.

Lippman, A. (1992). Led (astray) by genetic maps: The cartography of the human genome project and health care. *Social Science & Medicine, 35*(12), 1469–1476.

Lock, M. (2005). Eclipse of the gene and the return of divination. *Current Anthropology, 46*(S5), S47–S70.

Low, L., & King, S., et al. (1998). Genetic discrimination in life insurance: Empirical evidence from a cross sectional survey of genetic support groups in the United Kingdom. *British Medical Journal, 317*(7173), 1632–1635.

Mackintosh, M., & Parr, A. (2004). *New South Wales state and regional population projections 2001–2051.* Sydney, NSW: NSW Department of Infrastructure Planning & Natural Resources.

Marteau, T., & Richards, M. (Eds.). (1996). *The troubled helix: Social and psychological implications of the new human genetics.* Cambridge; New York: Cambridge University Press.

Mason, B., & Donnelly, P. (2000). Impact of a local newspaper campaign on the uptake of the measles mumps and rubella vaccine. *Journal of Epidemiology & Community Health, 54*(6), 473–474.

Mattick, J. (2003). Challenging the dogma: The hidden layer of non-protein-coding RNAs in complex organisms. *Bioessays, 25*(10), 930–939.

Martin, L. H., & Gutman, H., et al. (Eds.). (1988). *Technologies of the self: A seminar with Michel Foucault.* Amherst, MA: University of Massachusetts Press.

Mattick, J. S. (2004). The hidden genetic program of complex organisms. *Scientific American,* 61–67.

McCombs, M., & Gilbert, S. (1986). In J. Bryant & D. Zillman (Eds.), *News influence on our pictures of the world. Perspectives on media effects* (pp. 1–15). Hillsdale, NJ: Erlbaum.

Michie, S., & di Lorenzo, E., et al. (2004). Genetic information leaflets: Influencing attitudes towards genetic testing. *Genetics in Medicine, 6*(4), 219–225.

Milmo, C. (2002). The appliance of science: Cahal Milmo explains why it is unreasonable of scientists to expect the media to do their job for them. In *The Independent* (p. 13). London.

Moynihan, R., & Sweet, M. (2000). Medicine, the media and monetary interests: The need for transparency and professionalism. *Medical Journal of Australia, 173,* 631–634.

Nathanson, K. L., & Wooster, R., et al. (2001). Breast cancer genetics: What we know and what we need. *Nature Medicine, 7*(5), 552–556.

Nelkin, D., & Lindee, M. S. (1995). *The DNA mystique: The gene as a cultural icon.* New York: Freeman.

Nelkin, D., & Lindee, M. S. (1999). Good genes and bad genes: DNA in popular culture. In M. Fortun & E. Mendelsohn (Eds.), *The practices of human genetics* (pp. 155–167). Dordrecht, the Netherlands; Boston; London: Kluwer Academic Publishers.

Nelkin, D., & Tancredi, L. R. (1994). *Dangerous diagnostics: The social power of biological information.* Chicago: University of Chicago Press.

Nerlich, B., & Dingwall, R., et al. (2002). The book of life: How the completion of the human genome project was revealed to the public. *Health, 6*(4), 445–469.

Nielsen, A. C., (2001). *Australian TV trends 2001.* Sydney, NSW: AC Nielsen Media International.

Nisselle, A., & Delatycki, M., et al. (2004). Implementation of HaemScreen, a workplace-based genetic screening program for hemochromatosis. *Clinical Genetics, 65*(5), 358.

Nuffield Council on Bioethics. (1999). *Genetically modified crops: The ethical and social issues.* London: Nuffield Council on Bioethics.

Omenn, G. S. (2000). Genetics and public health: Historical perspectives and current challenges and opportunities. In M. J. Khoury, W. Burke, & E. J. Thomson (Eds.), *Genetics and public health in the 21st century: Using genetic information to improve health and prevent disease* (Vol. 40, pp. 25–44). Oxford: Oxford University Press.

Palmer, S. (2005). Ghost in the gene. *BBC Focus 2005, 156*, 52–55.

Pálsson, G., & Harðardóttir, K. E. (2002). For whom the cell tolls: Debates about biomedicine. *Current Anthropology, 43*(2), 271–301.

Petersen, A. (1998). The new genetics and the politics of public health. *Critical Public Health, 8*(1), 59–71.

Petersen, A. (2001). Biofantasies: Genetics and medicine in the print news media. *Social Science & Medicine, 52*(8), 1255–1268.

Petersen, A. (2002). The new genetics and the media. In A. Petersen & R. Bunton (Eds.), *The New Genetics and the Public's Health* (pp. 103–134). New York: Routledge.

Petersen, A. (2002). Replicating our bodies, losing our selves: News media portrayals of human cloning in the wake of Dolly. *Body and Society, 8*(4), 71–90.

Petersen, A. (2006). The best experts: The narratives of those who have a genetic condition. *Social Science & Medicine, 63*, 32–42.

Pirkis, J. E., & Burgess, P. M., et al. (2006). The relationship between media reporting of suicide and actual suicide in Australia. *Social Science & Medicine, 62*(11), 2874–2886.

Pitman, R., & Sanders, K., et al. (2002). Pilot study of secondary prevention of posttraumatic stress disorder with propranolol. *Biological Psychiatry, 51*(2), 189–192.

Polzer, J., & Mercer, S. L., et al. (2002). Blood is thicker than water: Genetic testing as citizenship through familial obligation and the management of risk. *Critical Public Health, 12*(2), 153–168.

Priest, S. H., & Talbert, J. (1994). Mass media and the ultimate technological fix: Newspaper coverage of biotechnology. *Southwestern Mass Communications Journal, 10*(1), 76–85.

Prior, L. (2001). Rationing through risk assessment in clinical genetics: All categories have wheels. *Sociology of Health & Illness, 23*(5), 570–593.

Rifkin, J. (1983). *Algeny*. New York: Viking.

Robotham, J. (2004). Genetic tests taken under pressure, survey finds. In *The Sydney Morning Herald* (p. 5). Sydney, NSW.

Robotham, J. (2006). Breast scan program deplored as coercion. In *The Sydney Morning Herald* (p. 6). Sydney, NSW.

Roper Starch. (1997). *Americans talk about science and medical news. A Roper Starch report on the council's survey: 21st century housecall: The link between medicine and the media*. Washington, DC: National Health Council.

Rorvik, D. M. (1978). *In his image: The cloning of a man*. London: Hamish Hamilton.

Rose, N. (2001). The politics of life itself. *Theory, Culture & Society, 18*(6), 1–30.

Rose, S. P., & Lewontin, R. C., et al. (1984). *Not in our genes: Biology, ideology, and human nature*. Harmondsworth, Middlesex: Penguin Books.

Roy Morgan Research. (2003). *Roy Morgan readership results for the year ending March 2003*. Melbourne, Victoria: Roy Morgan.

Ryan, C. (1991). *Prime time activism: Media strategies for grassroots organizing*. Boston: South End Press.

Scheufele, D. (1999). Framing as a theory of media effects. *Journal of Communication, 49*(1), 103–122.

Schmitz, D., & Wiesing, U. (2006). Just a family medical history? *British Medical Journal, 332*, 297–299.

Smith, D. P., & Armstrong, B. K. (1998). Prostate-specific antigen testing in Australia and association with prostate cancer incidence in New South Wales. *Medical Journal of Australia, 169*, 17–20.

Steele, E. J., & Lindley, R. A., et al. (1998). *Lamarck's signature: How retrogenes are changing Darwin's natural selection paradigm*. St Leonards, NSW: Allen & Unwin.

Turner, L. (1997). The media and the ethics of cloning. *Chronicle of Higher Education, 44*(5), B4–B5.

van Dijck, J. (1998). *Imagenation: Popular images of genetics*. Basingstoke, UK: Macmillan Press.

Van Hoyweghen, I., & Horstman, K., et al. (2006). Making the normal deviant: The introduction of predictive medicine in life insurance. *Social Science & Medicine, 63*(5), 1225–1235.

Voss, G. (2000). *Report to the human genetics commission on public attitudes to the uses of human genetic information. Human genetics commission*. London: The Human Genetics Commission.

Waalen, J., & Nordestgaard, B. G., et al. (2005). The penetrance of hereditary hemochromatosis. *Best Practice & Research Clinical Haematology, 18*(2), 203–220.

Wade, S., & Schramm, W. (1969). The mass media as sources of public affairs, science, and health knowledge. *Public Opinion Quarterly, 33*(2), 197–209.

Watson, J., & Crick, F. (1953). A structure for deoxyribonucleic acid. *Nature, 171*, 737.

Welsh Health Planning Forum. (1995). *Genomics: The impact of the new genetics on the NHS* (pp. 1–36). Cardiff: Welsh Office.

White, T. (1998). 'Get out of my lab, Lois!': In search of the media gene. In R. Hindmarsh, G. Lawrence, & J. Norton (Eds.), *Altered genes: Reconstructing nature: The debate* (pp. 24–36). St Leonards, NSW: Allen & Unwin.

Wilkie, T. (1993). *Perilous knowledge: The human genome project and its implications*. London: Faber and Faber.

Wilkie, T., & Graham, E. (1998). Power without responsibility: Media portrayals of Dolly and science. *Cambridge Quarterly of Healthcare Ethics, 7*(2), 150–159.

Williams-Jones, B. (2003). Where there's a web, there's a way: Commercial genetic testing and the internet. *Community Genetics, 6*(1), 46–57.

Willis, E. (1998). Public health, private genes: The social context of genetic biotechnologies. *Critical Public Health, 8*(2), 131.

Willis, E. (2002). Public health and the 'new' genetics: Balancing individual and collective outcomes. *Critical Public Health, 12*(2).

Catriona Bonfiglioli, Ph.D
Faculty of Arts and Social Sciences
University of Technology

JAYNE CLAPTON

7. ETHICS OF RE-MEMBERING AND REMEMBERING

Considering disability and biotechnology

ABSTRACT

Encompassed within forecasts offered by proponents of biotechnology about cures and control of disease and disability, are also predictions of an enhanced society. However, if the citizenship of our society is to be 're-membered' in the future with processes of elimination being employed on some potential lives deemed not worth living, deep ethical scrutiny is required.

Hence, this discussion contends that when considering ethical decisions impacting upon the membership of future societies, there also exists an imperative to seek insight and wisdom by looking to past actions. The experiences and narratives of many people with disability and their families attest to some of these past actions whereby agendas of elimination have been imposed by practices founded upon both moral and socio-political exclusion.

When considering 're-membering' in this context, a deliberation upon the imperative to remember will utilise the proximal text of a conference venue to contend that ethical reflection in this biotechnological era must embrace a process of anamnesis of past practices, rather than the more common approach of amnesia.

INTRODUCTION

Whenever science and technology have intersected throughout the history of humanity, human societies have also undergone challenge, crisis and change in terms of structure, membership and participation. Important questions always emerge when such intersections take place. These include: How is the membership of the society changed? Who are in and who are out? Who have knowledge and power and who do not? Who are the decision-makers and who are the silenced? Who are designated as typical citizens and who as atypical? Which lives flourish and who are denied opportunities?

For instance, in this contemporary era, societies are being presented with the promises and prosperity of various biotechnological advances, particularly those offered in the fields of biomolecular science and genetic technologies. Ethical reflection about how such technologies provide the capacity to redefine citizenship along with the measure of their humaneness for all persons is continually being deliberated upon. The goal of enhancing societies without causing harm to the members, or potential members, must be of prime ethical consideration.

Naomi Sunderland et al. (eds.), Towards Humane Technologies: Biotechnology,
New Media and Ethics, 97–112.

Such ethical consideration was the topic of discussion at an international conference held in Ipswich, Queensland, Australia, in July 2002 . At this event, constituent issues of biotechnology, the new media and citizenship were discussed by the conference participants. As part of this process, participants were also invited to consider possible 'sites of contestation' of the dominant processes for ethical scrutiny that are most often presented in this context. One 'site of contestation' was identified to explore the impact of various contemporary biotechnologies on people with disability. This site is particularly significant when consideration is given to historical accounts of the impact of technological advances in relation to claims of cures and control of disease and disability.

This discussion, which is adapted from a paper presented at the conference, contends that when pondering upon how the citizenship of our society may be 're-membered' in the future, as a consequence of biotechnology, ethical considerations that embrace the histories of people with disability become critical. By exploring narrative accounts of experiences of such people as recorded by family members, historical incidents where some ways of being human have been preferred to others are highlighted; and the often presumed foundations on which contemporary biotechnological practices are founded can be challenged.

Various steps are undertaken within the discussion. Why it is important 'to remember' is explored in a context of potential 're-membering'. The particularities of the context as experienced by some people with disability are presented against the theoretical explanations of the need to remember. Links between acts of re-membering and remembering in this biotechnological era are made when certain histories are highlighted. The ethical implications of such explorations are then discussed, suggesting that contemporary practitioners become more aware that ethical deliberations should be performed in a context of anamnesis - the act of remembering or recollecting, rather than amnesia – the act of not-remembering.

WHY REMEMBER?

Two reasons can be identified for why the conference should have considered the interaction between ethics, biotechnology and disability as a 'site of contestation'.

Firstly, for people with disability, ethical implications have often left a legacy of not just moral exclusion, but of moral betrayal. In ethical deliberations within post-Enlightenment, industrialised Western societies, a particular type of citizen is seemingly more morally privileged than other types. When ethical deliberations depend upon the qualities of reason, autonomy and independence, those deemed without these qualities are continuously at risk of being pushed to the margins and beyond. Presumptions of both diminished personhood and indeed questionable humanity therefore conjointly embrace the tendency to both disqualify such beings from moral agency and to render their claims to moral membership and protection as somewhat irrelevant (Clapton, 2000, 1999).

The second reason was provided by the physical site of the conference venue. The Ipswich Campus of Queensland University presents as a contemporary campus of a modern university whereby the aesthetics of new buildings fuse with the heritage architecture of a time gone by when the site was used for other purposes.

For over 100 years, such a purpose was that of being a large institution for both people with psychiatric disability and those with intellectual disability. The physical site represented, then, a proximal 'text' of an historical account of technologies of cure and control in relation to people with disability.

It seems apparent that in order to consider predictions about future humanity as proposed by biotechnological advances, there is an ethical imperative to seek insight and wisdom by looking to the past. In doing this, we are challenged to reflect upon why ethical agendas about the future should not be propelled from a position of amnesia of past events, but rather from anamnesis.

French philosopher, Paul Ricoeur (1999, p. 5) states that remembering is a kind of action linked to exercising memory not just with words, but also with our minds. Furthermore, he describes three levels of practical applications of evoking memories for ethical significance: the pathological–therapeutic level, the pragmatic level, and the ethical-political level (Ricoeur, 1999, p. 6-7). Whilst the pathological–therapeutic level explores the wounds and scars of memory, along with notions of mourning and melancholia linked to the work of memory, it is the second and third levels that are pertinent to this discussion.

Ricoeur (1999, pp. 7-9) describes how the pragmatic level represents the praxis of memory. It is this level in which problems of identity are explored. Whether personal or collective, the identity question of 'who am I', rather than the attributed 'what am I?' or 'what are you?' is constituted both by interactions with others and experiences of the self in time. He proposes that such identities are thus articulated through narrative, and represent a rich source of ethical insight. The significance of the narratives becomes apparent not just in the expression of life experiences of the past, but also in the relevance for the future. It is the relationship between such experiences and ethical principles such as justice, beneficence and non-malevolence that are enlightening. Inasmuch as what is told or untold, what plot devices are used or not used, and what exemplars of events are discussed or not discussed, narratives are thus constituted and a subsequent form of ethical reflection becomes available.

Ricoeur (1999, pp. 9-11) declares that in remembering, we are not only showing deep concern for the past, but are also being bound to transmit that past to future generations. Therefore, the ethical-political level of the application of memory considers whether there is, in fact, a 'duty to remember'? In a context where history tends to celebrate victors and privilege the accounts of the powerful, Ricoeur (1999, p. 11) contends that there is a duty to remember to keep alive the memory of suffering.

Through the use of narrative, the context of re-membering in this biotechnological era, presents a critical ethical opportunity. In fact, according to Saul (2001, p. 215), not to remember will only create a vacuum in which propaganda and ideology of dominant agendas thrive. If the voices of those who have previously suffered at the hands of dominant proponents remain silent and their narratives never sought, the expression of the identity question 'who am I?' is foreclosed. Such vulnerable people are constantly at risk of only being determined and referred to by the powerful in the objectified terms of 'what are they?'

Nelson's (2001) description of this phenomenon offers an important insight for the moral implications.

Identities mark certain people as candidates for certain treatments, and within abusive group relations these treatments are seldom benign. The connection between identity and agency poses a serious problem when the members of a particular social group are compelled by the forces circulating in an abusive power system to bear the morally degrading identities required by that system. (Nelson, 2001, pp. xi–xii)

A sustained suppression through not remembering also has an important moral effect. It prevents the opportunity for counterstories to be offered by such vulnerable identities in order to both subvert master narratives of control and to offer deeper ethical insight for the future (Nelson, 2001, p.69ff). Hence, the ethical relationship between biotechnology and disability must encompass not only the scientific capacities to re-member, but also the narratives of people with disability themselves or of their families, advocates/carers as positive counterstories to the memories of past atrocities.

RE-MEMBERING AND REMEMBERING

Utilising a broad range of technologies and possibilities, biotechnology promises to open up many of the unsolved medical 'problems' facing humanity. Paradoxically, though, such a pursuit is predominantly undertaken by a reductive project of modern biology, supported by strongly reductionist and analytical philosophical programs that have several implications for modern life sciences and humanistic relations (Sloan, 2000, p. 9). According to Sloan (2000, p. xxiii), biotechnology poses questions about the nature of the human person, about the moral limitations on the technological manipulation of life, and about the definition of individual and social goods. If understandings of humanity are thus defined and causally determined within biotechnological contexts by micro parts, rather than expansive metaphysics, an ongoing ethical challenge remains. To what extent, then, are identity questions of 'who we are', rather than merely 'what we are' intrinsic to ethical discussions.

Assumptions of what it means to be human, or more pertinently perhaps, what constitutes the right way of being human are also implicit within such contexts. Individuals, or potential individuals, who fall into a biologically determined category of 'not-right' are at risk of their human membership being denied; and their personhood, or potential personhood, being negated. In other words, in such a biologically and philosophically determined reductionist context, the anomalous Other to the prototypical understanding of being human is at risk of elimination (Clapton, 1999). They are the negative, the bad, the not-desired, the harmful and the hunted in a context signified by enhancement, doing good and victory. With profound similarities to historical processes of colonisation (Clapton, 2004; Sibley 1995), an agenda of choice and control becomes driven by an imperialist quest to conquer the unknown and overcome inferiority or imperfection (Dreger, 2000, p. 159ff). New technologies involving genetic testing, screening and diagnosis, advanced Assisted Reproductive Technologies, and improved detection and intervention procedures now impact upon which 'humans' will constitute the membership of contemporary and future societies, and which humans will be denied life on the basis of being deemed a life not worth living.

People with disability become enmeshed in a web of constructions overseen by powerful disciplines such as science, medicine, law, philosophy and politics. Their status is viewed as representative of inferior or flawed humanity, but their presence is nonetheless respected. Proponents of biotechnology do not promote the elimination of living people with disability. Instead though, how such people's lives are interpreted, does inform choices made in relation to procreative choices, foetal diagnosis and other situations involving imperilled or 'needy' humans.

There is no denying that advances in science and medicine have had many positive effects for the well being of humanity, whether such humans have disability or not. Biotechnological advances are thus promoted not so much in a context of control, but rather of care. Choices about which humans should live are not overtly made on the basis of elimination of inferiority, but rather within concerns of care and compassion towards the relief of suffering and burden. If people with disability are perceived as dependent, useless, harm-causing and non-productive by dominant moral subjects, then an ethical imperative of biotechnological practitioners to relieve or prevent the tragedy of their presence is seemingly justified. Consequently, if such practices also result in acts of elimination, these can be merely defended by a notion of 'double effect' – an effect by which the salvific intention also has political consequences of sustaining a particular social order which privileges superiority. These constructions, which constitute both a personal tragedy view of disability along with socio-political effects resulting in exclusion and oppression, are well documented in contemporary disability studies literature (see for example, Barnes and Mercer, 2003; Barnes, Oliver and Barton, 2002).

It is not surprising that historical links to the promotion of personal tragedy approaches continue to underpin contemporary biotechnological practices. Consequently, the capacity to determine which potential humans should be given life, is also often situated within the realm of medical ethics, and comfortably fits within these approaches (Clapton, 2003).

The ethical tenets associated with this 'tragedy' context, however, are becoming increasingly debated. Disability rights activists and other writers argue that bio-technology allows for discriminatory choices made against people with disability, and that such choices have their foundations in eugenic practices (see for example, Rowland, 2001; Wolbring, 2001; Kitcher, 2000; Vehmas, 1999; Shakespeare, 1998; and Hubbard, 1997).

Whilst ignoring the theoretical embeddedness of their stance, opponents of this view powerfully respond that any talk of active discrimination and eugenics within biotechnological practices is unfounded. Instead, they argue that the procreative choices of parents to not have a child with a disability should be respected (see for example, Gillon, 2001; Gillott, 2001; Harris, 2001, 2000 and 1993). Other writers recognise that some choices may indeed be eugenic, but again they are only made within the domain of personal choice, and not because of notions of political force, coercion, compulsion or threat towards particular populations (Caplan, 2000, pp.

218-222). Savulescu (2002, np.) does concede, though, that when testing and termination are the only choices presented in pregnancies where Down Syndrome is detected, for instance, unjustifiable eugenics and discrimination does take place.

Koch (2001, np.), himself a person with a disability, suggests that in order to consider such deep ethical debates which perceive disability as inherently harmful and negative, ethicists, scientists and medical practitioners must consider insider voices and narratives of those who can relate the 'experience' of disability, rather than ethical deliberations being made on the basis of biological determinism alone. However, this approach does not readily fit with the dominant ethics employed in such contexts. Practices of care and control, embedded in both duty-based and utilitarian ethics (Vehmas, 1999: Sobsey, 1994), are easily linked to a view of people with disability as objects of science and medicine, and as targets of ethical action which require amelioration, prevention or cure.

Arguably, though, what is missed in the contemporary discussions of biotechnology, disability and ethics, is not whether contemporary and future practices are eugenic, but rather what constitutes past patterns of the experiences of people with disability with practices of care and control. The suspicion that is evoked from these experiences is pertinent and is particularly relevant when considering past experiences of people with intellectual and/or psychiatric disability. Past (and present) practices of care and control have not only denied a sense of belonging for people with disability within society, but they have also concealed or prohibited the telling of their narratives. In a context where individuals' intellectual capacities may already constrain the possibilities of narratives, suspicions increase when narratives that can be told are not only ignored nor sought, but also when the experiences within the narratives are depictions of profound exclusion, violence and abuse (Clapton , 2000, 1999).

In a biotechnological era of re-membering the future, the imperative to remember past practices towards people with disability is apparent. When these ascribed anomalous Others have suffered at the hands of privileged practices and ideologies of dominant disciplines, we also have a duty to remember so that their experiences will not be forgotten when similar patterns present themselves before humanity.

REMEMBERING THE SITE

Saul (2001, p. 247) states, "History is the story of memory sorting itself out. Events and people are retained or dropped." Therefore, it seems, the integrity of the history is located in the direct relationship between the tellers, what is presented, and what has been deemed important to present. The proximal 'text' of the conference venue, the Ipswich Campus of the University of Queensland, presents an exemplary context for remembering. However, remembering needs to account for three presenting histories: the recorded history, the analysed history and the experiential history. The use of narratives provides the opportunities for painful stories concerning people with disability to reflect how hegemonic interpretations of the ethical principles of autonomy, justice, beneficence and non-malevolence have not only affected their lives, but have sustained their vulnerability, particularly in regard to societal membership.

THE RECORDED HISTORY

'The past' of this text is presented in the UQ Ipswich Campus Guide (received 2002) as:

> The UQ Ipswich Campus stands on a site once occupied by the Sandy Gallop Asylum, a branch of the Goodna Asylum, which began operations in 1878, housing 50 mentally ill patients. Sandy Gallop can be described as "the most complete example of an asylum in Queensland based on the principles of 'moral treatment'".

Moral treatment emphasised a pleasant environment for patients, and at Sandy Gallop included employment and recreation areas, well-designed buildings, gardens and sweeping views of the countryside. (To preserve these views, the asylum employed sunken fences known as "ha-has", once common on English estates.)

In 1968, the site was renamed the "Challinor Centre" after Dr Henry Challinor, a prominent local doctor and Queensland Parliamentarian of the mid-19th century. Reflecting changing practices in the care of people with intellectual disabilities, the Challinor Centre operated until August 1998, when the Centre was closed and construction of stage one of UQ Ipswich began.

The brochure refers to practices by which people with mental illness, psychiatric disability and intellectual disability were congregated together in custodial care in Queensland up until the late-twentieth century. Until the institution closed in 1998, there were many twists and turns in its history. As recorded by McRobert (1997), the notion of a complete institution was a facility that not only contained 'patient' and staff residential wings, but was also self-sufficient in all auxiliary aspects. For instance, the production and preparation of food, the doing of laundry, the maintenance of the buildings, the venues for recreation – including a cinema, the inclusion of a surgical building and operating theatres, as well as the presence of a mortuary and post-mortem facilities were all located within the site. In other words, all provisions were attained within the confines of the institution.

THE ANALYSED HISTORY

In social terms, post-Reformation and newly industrialising societies became focused on productivity and profitability, underpinned by the forces of developing capitalism and the Protestant Work Ethic. In the second half of the eighteenth century the changing structure of the English economy underpinned structural changes in relationships of social order, such as the transition from master-servant, to employer-employee - relationships now dominated by rank, order, and class (Scull, 1993, p. 31). People with disability became a 'class' requiring physical removal from the 'able-bodied' norms of mainstream society. The 'institution' or 'asylum', involving the State, evolved as a place with a dual purpose: (a) where people with disability could be placed so that family members could meet workers' obligations; and (b) a place where people with disability could be skilled to become productive members of society (Funk, 1987).

The history of the Challinor Centre refers to the influence of moral treatment and also to the work of its American proponents, Dorethea Dix and Samuel Gridley Howe who promoted a view of the asylum as a place of refuge. However, being established as a 'lunatic asylum' also locates it within a particular historical context. The 'lunatic asylum' was thus defined as both an institution for the support, safekeeping, cure or education of those incapable of caring for themselves such as people with mental illnesses or disabilities; and as benevolent because it supported people with insufficient income (McRobert, 1997, p. 5).

Sandy Gallop was opened at a time which coincided with the impact of legislation passed to restructure the English and Welsh 'mad-business' in mid-19th century Britain. This shift to a formalised, medicalised context also saw the ascendancy of the profession of the so-called 'mad-doctors' (Philo, 1987, p.398). Therefore emerging practices of diagnosing, prognosing, categorising and pathologising those with mental illness and intellectual disability within asylums were situated in a professional context in which status and recognition were being sought. For instance, Philo (1987, p. 399) notes that in the first forty issues of the Asylum Journal established in the mid-19th century by the authority of the Association of Medical Officers of Asylums and Hospitals of the Insane, the profession was seeking to establish its special expertise within the public arena. In comparison to previous practices offered by lay people who had previously been responsible for 'institutionalised lunatics', this new breed of professionals broadcast their self-proclaimed, salvific status:

> "Oh what a holy, honourable and sacred occupation is that which we all have the privilege to be engaged: the angels of heaven might well envy us the ennobling and exalting pleasures incidental to our mission of love and charity."

> "The physician is now the responsible guardian of the lunatic and must ever remain so … insanity lies strictly within the domain of medical inquiry." (Philo, 1987, p. 400)

The practices that developed in these asylums around notions of care and control towards anomalous Others, became embedded in a medico-moral discourse which represented 'medico-moral treatments'. These had some particular characteristics, which were not only clinical, but also physical and geographical; and which required spatial separation. Philo (1987, p.404) describes how there was a prevailing view in this Asylum era "that the mentally distressed mind could only be cured by freeing it from the city and the factory, and by then giving it the benefits of a more "natural" tranquil rural setting."

Sustaining the supposedly beneficent telos of institutional 'care', however, became an enduring tension. The intention to skill people to be productive and returned to society could not only be unrealised in many situations, but also proved to have a limited vision. The institutionalisation trend occurred contemporaneously with the beginning of the modern era, with an increasing emphasis on scientism and social Darwinism, and an ongoing categorisation and classification of different humans. Social Darwinism would have a profound affect, not only for Darwin's notion of the "evolutionary advantage of the fittest", but also for providing the foundation by which his cousin, Francis Galton, would develop the practice of Eugenics, the implications of which are highlighted by Davis (1995).

> Darwin's ideas serve to place disabled people along the wayside as evolutionary defectives to be surpassed by natural selection. So, eugenics became obsessed with the elimination of 'defectives,' a category which included the feeble-minded,' the deaf, the blind, the physically defective, and so on. (Davis, 1995, pp. 30-31)

The advent of statistical science was another great influence that would have deadly results. The eugenics movement progressed as an application of the development of statistics and the construction of the Normal curve; and hence was linked to an ongoing construction of normalcy (see for example, Davis, 1995, Trent, 1994). This is a complex concept which shifted from understanding 'typical' humans from a physiological view; to defining 'right' humans according to statistical and psychometric means. Along with the study of statistics, 'normalcy' became not only the construction of excellence, but, considering the underside, also the construction of mediocrity and deficiency (Davis, 1995).

Promoted by prominent physicians and scientists of the late 1800's, eugenics is the 'pseudo- science' that dealt "with the improvement (as by control of human mating) of the hereditary qualities of a race or breed." (Sobsey, 1994, p.119) Whilst the Darwinian view of natural selection depended upon nature's actions to eradicate, the eugenic view attempts to "defy nature", in favour of actions by a "privileged class who exercises control over the rest of humanity" by the use of technologies to manipulate nature (Sobsey, 1994, p. 120). Once people recognised the power in statistically ranking human superiority, the idea of increasing intelligence of humans and decreasing birth defects became achievable (Davis, 1995, p. 29ff). These interests were underpinned by the use of the body metaphor in reflecting the state of the nation and national fitness. If individual citizens are not fit, then the national body will not be fit (Davis, 1995, p. 36). For people with disability, physical institutions became a site of segregation to both house the 'scientifically-determined' inferior humans and to prevent procreation. Wendell highlights how in cultures supported by modern Western medicine, and which idealise the idea that the body can be objectified and controlled, "those who cannot control their bodies are seen as failures." (Wendell, 1992, p. 72)

The history of institutionalised practices, therefore, is both complex and ambiguous. When anomalous Others have interacted with dominant societal practices of an era, they have been subjected to various representations of 'care' and 'control'. When these practices also encompass practices of elimination and undesirability, they have been underpinned by dominating disciplines like medicine, science, politics and philosophy which have been controlling in their determination of what represents burden and non-productivity. Whatever the context, when anomalous Others are configured by these disciplines as dependent and suffering and in need of professional determination and intervention, this has also significantly sustained an inferior moral status being applied to such beings or potential beings (Clapton, 1999).

THE EXPERIENTIAL HISTORY

Institutional practices progressed until late in the second half of the 20th century. Children born with different conditions were relinquished by many families and placed in institutional care on the advice of medical personnel who declared to parents that such children could not be cared for at home because of the burden of care. With no other forms of support available in the community, families not only reluctantly surrendered their role as primary caregivers, but also felt a complex set of emotions with the deep sense of loss of a loved one. This was indeed the experience of quite a number of families of Challinor residents.

Narratives of some of these families have been recorded by the Community Resource Unit (CRU) in Queensland. The tenet of the narratives can be summed up in the comments of one parent, "The institution was a horrible experience. Nobody should have to go through that." (CRU, 2001: p72)

Other themes that become apparent are: despair; denial; separation; powerlessness; pain; confusion; obscenity; violence; horror and control. It is useful to look at some examples.

Despair

Parent:

> "Parting with you when you were still a child was difficult for both of us. You fretted and so did I. Many times I thought my tears would never cease, especially when I saw other children laughing and playing. I have often wondered what your thoughts were. You have never said. I wonder if I shall ever know your side of your life-story." (CRU, 2001,: p49)

Denial

Family member:

> "The Managers were defenders of the system. We had the feeling that their main concern was to ensure that things got covered up: The truth mustn't come out. When complaints were made it was the workers who got counselled. The people who had suffered at their hands were left to cope with it by themselves." (CRU, 2001: p70)

Separation

Mother:

> "I look back over the seven years in the institution for Lynette. Gone are those adolescent years. What was lacking was our participating role as parents. We should have been involved in her life." (CRU, 2001: p65)

Powerlessness

Sister:

"They even changed Renee's name, although not deliberately, but it is an example of how they took control without consideration for the family. For almost forty years Renee was known as 'Reenie' at Challinor. When we would ask about Renee, staff looked at us blankly until we also referred to Renee as 'Reenie'." (CRU, 2001: p75)

Pain

Mother:

"My new baby was sick and I couldn't cope. Mum and Pop said it was the hardest thing they ever had to do, to take Annie to the Centre and leave her there. Pop never got over it. Mum accepted it. Pop was working each day and I don't think he realised how hard it was for me to manage." (CRU, 2001: p35).

Confusion

Parent:

"No one would ever tell me if Libby was ever going to get better. Doctors didn't talk to you in those days and I didn't know what questions to ask. I was so naïve. When Libby was three we took her to a specialist at the Mater Hospital. He suggested that we should put her away in a home. He said she would ruin our lives. He didn't tell me why he thought that, he just said, "Look, you'll be better off putting her into an institution.". When I think back I still feel very upset about those comments." (CRU, 2001: p67)

Obscenity

Mother:

"Accompanied by the social worker, Libby and I arrived at Challinor and were taken to Francis House. This was the worst area of all at Challinor. We went in through the side door and the first thing that hit me was the stench of urine in the cracked tiles. The beds were all lined up. I remember a huge nursing sister, so large she could hardly walk. She showed me the solarium where Libby would be during the day. It was absolutely gross. Lots of people were sitting around the walls, propped up, in their own urine. I saw a mop in the corner. I'll never forget seeing someone pick up this stinking mop and mop up all the urine, but actually making it worse. There were faeces everywhere. Some people weren't even wearing pants." (CRU, 2001: p44)

Violence

Sister:

> "I heard a whack and a scream…. I knew it was someone being hit. I turned to my mother and said, "Do they do that to Timmy, Mum?" She just stood there looking at me. I remember the look on her face and the tears in her eyes. She didn't know what to say to me. I will never know the pain and heartache my parents endured back then." (CRU, 2001: p55)

Horror

Family members:

> "No words can really describe that place. It was so awful."

> "It was a place of detention, a place of containment."

> "It would shock even the most hardened person." (CRU, 2001: p72)

The narratives all record an ambiguity experienced by families in recognising that the place offered both site of care and refuge as well as helplessness and entrapment. This form of ambiguous memory is recorded by a sister:

> "We were then escorted through smelly, noisy, dank wards, past people in cages or rolling around on concrete floors, to a room where Renee had been placed for our visit, and the room was then locked. Renee never smelt nice. There was strong smell of lots of Johnson's baby powder, but always the lingering hint of urine and faeces, the stale smell of soap, the odour of neglected teeth. … not the stuff of fond memories." (CRU, 2001: p54)

REMEMBERING THE ETHICS

As Saul (2001, p. 257) states,

> … [M]emory moves in cycles and circles and so overtake us when least expected. We are then caught in the repercussions of our actions. Suddenly we recognize the parallels with past acts. We both remember and are powerless to use that remembrance. Only a circular or cyclical approach would allow us to act and remember at the same time.

In line with Ricoeur's (1999, pp. 9-11) notion of pragmatic memory, traumatic narratives exemplify the many forms of past atrocities. Such remembering is not only to respect and recognise key identity markers of individuals, but arguably, also the collective identity of humanity in terms of how we interact with an Other. When groups of people such as people with disability experience both moral and socio-political exclusion, Ricoeur's (1999, pp. 9-11) notion of ethical-political memory is also pertinent. Physical sites representing practices of such exclusion are ethically informing in regard to contemporary actions.

Why should we remember these histories in this contemporary era? What relevance do they have in discussions about biotechnological practices?

On the whole, the memories within these narratives and histories are not ones of humane treatment. The narratives in particular, exemplify the dominant experiences of anomalous Others as prescribed by practitioners of powerful disciplines with a capacity to control destinies. People with disability, when confronted with agendas of elimination or separation underpinned by notions of undesirability and burden experience oppression, cruelty and societal exclusion. Arguably, then, if ethics is a discipline to understand good and bad, right and wrong, justice, beneficence and non-malevolence, then the failure to protect such vulnerable Others on the basis that they represent not right ways of being human must constitute moral betrayal. If the presence of disability is constructed ethically as representing harm, threat and unhappiness within discourses of tragedy and catastrophe, utilitarian ethics thrive, and underpin practices of domination and prevention (Clapton, 2003). When considering and ethically scrutinising the biotechnological agenda of re-membering the future one must be insightful to parallel patterns within the past.

Contemporary biotechnological practices provide mechanisms seemingly to prevent anomalous Others being subjected to the presumed trauma of having a disability. If such potential suffering, as well as economic and care-giving burden, can be prevented by technologies of detection and eliminating intervention, then according to utilitarians such as Harris (2000), this represents a good moral action. Of particular significance is that Harris (2000, np) then goes further to claim that to knowingly fail to avail oneself of such possibilities would in fact be immoral.

In the 21st century, the view that it is morally right or good to prevent the life of a person deemed to potentially suffer, is problematic. It merely represents a now somewhat outdated view of disability as a personal problem or tragedy of an individual or a family This discussion has indicated, though, that what people suffer is not so much linked to their particular diagnosed conditions, but rather to how 'different' humans treat each other. How humanity and society is shaped according to science, medicine, philosophy, politics and economics affects such interactions. The narratives of representative groups who are rendered vulnerable by such deliberations are ethically important.

Ethical deliberations in regard to biotechnological implications for contemporary and future citizenship, however, should not be constrained to only narratives of moral and social exclusion. Other narratives representing what Nelson (2001) refers to as 'counterstories' can also be found. There is a significant body of literature supporting positive experiences of families who have a member with disability (for example, see Kittay with Kittay, 2000; Stainton and Besser, 1998).

How the focus of ethics towards people with disability in this biotechnological era comprehends relational acceptance, mutuality and interdependence found within counterstories, will challenge the dominant moral assertions promoted by disciplinary 'experts'. Burden, in the context of counterstories, is recognised not as that configured by judgements of individual limit or loss ascribed according to a diagnosis, but rather that which is due to the lack of opportunities, support and a sense of integral belonging as 'offered' by an excluding society (Dowling and Dolan 2001; Clapton, 1999; Kittay, 1999; Landsman, 1998).

It becomes apparent that powerful disciplines involved in contemporary biotechnological practices need to revise their moral standpoints on these matters to include the moral knowledge gleaned from counterstories. Hence, contemporary views and scholarship which critique the long established personal tragedy approaches about disability are more useful to inform public discussions than those that continue to perpetrate moral and social exclusion. Not to do so will see disability activists continue to cast the actors of these disciplines who can only understand disability in negative, deficit-focused ways, as villainous, and the opportunity for deep ethical reflection as prohibitive

CONCLUSION: REMEMBERING BEFORE RE-MEMBERING

In conclusion, this discussion has indicated how our present society must attend to the ethical challenges that emerge when we engage with the pragmatic and ethical-political memories available.

Saul (2001, p. 254) asks whether memory should have an object; and therefore whether the act of remembrance should provide access to that object? This discussion has also highlighted how the Ipswich Campus of the University of Queensland provides an object by which to remember some of the lived experience of people subjected to agendas of elimination, separation and undesirability. Accounts of practices of care and control reveal an ambiguous context. Rather than compassion and protection, such experiences reflect tragic notions of oppression, violence and exclusion. When ethics, as a discipline, deliberates upon the capacities for biotechnological manipulations to determine which humans can be members of society, ethical reflection will be constrained and moral and social exclusion perpetuated if such deliberations are based only on the 'presumed' tragedy of individual lives rather than the tragedies imposed by inadequate social supports. Hence, the tragic implications of inhumane practices towards vulnerable, anomalous Others must be remembered not only to continually remind us of past examples, but also to prevent ethical deliberations being based upon now somewhat outdated premises.

When affected groups do not have access to ethical decision making, when their voices are not sought or are denied, and when sites of practices are converted and sanitised for other purposes, we lose, through these amnesic practices, the moral importance of the pragmatic and ethical-political memories that they offer. It is hoped, then, that in this biotechnological era requiring ethical reflection, we, as a collective society, will have the moral insight that such a loss be remedied. When processes of anamnesis are enacted and counterstories of positive experiences of disability are embraced, disability as a 'site of contestation' could be ethically refigured and transformed.

REFERENCES

Barnes, C., & Mercer, G. (2003). *Disability*. Cambridge: Polity.

Barnes, C., Oliver, M., & Barton, L. (Eds.). (2002). *Disability studies today*. Cambridge: Polity.

Caplan, A. (2000). What's morally wrong with eugenics? In P. Sloan (Ed.), *Controlling our destinies: Historical, philosophical, ethical, and theological perspectives on the Human Genome Project*. Notre Dame, IN: University of Notre Dame Press.

Clapton, J. (2004). Disability, Ethics and Biotechnology: Where are we now? In C. Newell & A. Calder (Eds.) Voices in Disability and Spirituality from the Land Down Under (pp. 21-31). New York: The Haworth Pastoral Press.

Clapton, J. (2003). Tragedy and Catastrophe: Contentious discourses of ethics and disability. Journal of Intellectual Disability Research (Special Issue: Ethics and Intellectual Disability), 47 (7), pp. 540-547.

Clapton, J. (2000). Irrelevance personified: An encounter between disability and bioethics. *Interaction*, *13*(4), 11–15.

Clapton, J. (1999). *A transformatory ethic of inclusion: Rupturing 'disability' and 'inclusion' for integrality*. Unpublished PhD dissertation, Queensland University of Technology, Brisbane.

Community Resource Unit (Written by B. Funnell). (2001). *Telling the untold: Families, disability and institutions – stories of banned and unrequited love*. Brisbane: CRU Publications. (narratives reprinted with permission of CRU)

Davis, L. (1995). *Enforcing normalcy: Disability, deafness and the body*. London: Verso.

Dowling, M., & Dolan, L. (2001). Families and children with disabilities – IN equalities and the social model. *Disability & Society*, *16*(1), 21–35.

Dreger, A. (2000) Metaphors of morality in the human genome project. In P. Sloan (Ed.), *Controlling our destinies: Historical, philosophical, ethical, and theological perspectives on the human genome project*. Notre Dame, IN: University of Notre Dame Press.

Funk, R. (1987). Disability and rights: From caste to class in the context of civil rights. In A. Gartner & T. Joe (Eds.), *Images of the disabled: Disabling images*. New York: Praeger.

Gillon, R. (2001). Is there a 'new ethics of abortion'? *Journal of Medical Ethics*, *27*(Suppl.), 115–119.

Gillott, J. (2001). Screening for disability: A eugenic pursuit? *Journal of Medical Ethics*, *27*(Suppl.), II 121–II 123.

Harris, J. (2001). One principle and three fallacies of disability studies. *Journal of Medical Ethics*, *27*(6), 383–387. (Electronic version)

Harris, J. (2000). Is there a coherent social conception of disability? *Journal of Medical Ethics*, *26*(2), 95–100. (Electronic version)

Harris, J. (1993). Is gene therapy a form of eugenics? *Bioethics*, *7*(2/3), 178–187.

Hubbard, R. (1997). Abortion and disability: Who should and should not inhabit the world? In L. Davis (Ed.), *The disability studies reader*. New York: Routledge.

Kitcher, P. (2000). Utopian eugenics and social inequality. In P. Sloan (Ed.), *Controlling our destinies: Historical, philosophical, ethical, and theological perspectives on the human genome project*. Notre Dame, IN: University of Notre Dame Press.

Kittay, E., & Kittay, L. (2000). On the expressivity and ethics of selective abortion for disability: Conversations with my son. In E. Parens & A. Asch (Eds.), *Prenatal testing and disability rights*. Washington, DC: Georgetown University Press.

Kittay, E. (1999). *Love's labour: Essays on women, equality, and dependency*. New York: Routledge.

Koch, T. (2001). Disability and difference: Balancing social and physical constructions. *Journal of Medical Ethics*, *27*(6), 370–376. (Electronic version)

Landsman, G. (1998). Reconstructing motherhood in the age of "perfect" babies: Mothers of infants and toddlers with disabilities. *Journal of Women in Culture and Society*, *24*(1), 69–99.

McRobert, E. (1997). *Challinor centre: The end of the line – A history of the institution also known as Sandy Gallop*. Brisbane: Queensland Government.

Nelson, H. L. (2001). *Damaged identities: Narrative repair*. Ithaca, NY: Cornell University Press.

Philos, C. (1987). "Fit localities for an asylum": The historical geography of the nineteenth-century "mad-business" in England as viewed through the pages of the Asylum Journal. *Journal of Historical Geography*, *13*(4), 398–415.

Ricoeur, P. (1999). Memory and Forgetting. In R. Kearney & M. Dooley (Eds.), *Questioning ethics: Contemporary debates in philosophy*. London: Routledge.

Rowland, R. (2001). The quality control of human life: Masculine science & genetic engineering. In R. Hindmarsh & G. Lawrence (Eds.), *Altered genes II: The future*. Melbourne: Sage Publications.

Saul, J. R. (2001). *On equilibrium*. Ringwood, Victoria: Penguin Books.

Savulescu, J. (2002). Is there a "right not to be born'? Reproductive decision-making, options and the right to information. *Journal of Medical Ethics*, 28(2), 65–67. (Electronic version).

Scull, A. (1993). The most solitary of afflictions: Madness and society in Britain 1700–1900. New Haven, CT: Yale University Press.

Shakespeare, T. (1998). Choices and rights: Eugenics, genetics and disability equality. *Disability & Society*, 13(5), 665–681.

Sibley, D. (1995) *Geographies of exclusion: Society and difference in the West*. Routledge: London.

Sloan, P. (2000). Completing the tree of Descartes. In P. Sloan (Ed.), *Controlling our destinies: Historical, philosophical, ethical, and theological perspectives on the human genome project*. Notre Dame, IN: University of Notre Dame Press.

Sobsey, D. (1994). *Violence and abuse in the lives of people with disabilities: The end of silent acceptance*. Baltimore: Paul Brookes Publishing Co.

Stainton, T., & Besser, H. (1998). The positive impact of children with an intellectual disability on the family. *Journal of Intellectual Development and Disability*, 23(1), 57–70.

Trent, J. (1995). *Inventing the feeble mind: A history of mental retardation in the United States*. Berkeley, CA: University of California Press.

Vehmas, S. (1999). Discriminative assumptions of utilitarian bioethics regarding individuals with intellectual disabilities. *Disability & Society*, 14(1), 37–52.

Wendell, S. (1992). Towards a feminist theory of disability. In H. Holmes & L. Purdy (Eds.), *Feminist perspectives in medical ethics*. Indianapolis, IN: Indiana University Press.

Wolbring, G. (2001). Where do we draw the line? Surviving genetics in a technological world. In M. Priestley (Ed.), *Disability and the life course: Global perspectives*. Cambridge: Cambridge University Press.

Jayne Clapton PhD
School of Human Services
Griffith Abilities Research Program
Griffith University

ELEANOR MILLIGAN

8. GENETIC SCREENING AND PRENATAL TESTING

"Women are situated on the research frontier of the expanding capacity for prenatal genetic diagnosis, forced to judge the quality of their own foetuses, making concrete and embodied decisions about the standards of entry into the human community" (Rapp, 2000, .p. 3).

"In this liberal and individualistic society, there may be no need for eugenic legislation. Physicians and scientists need merely to provide the techniques that make individual women, and parents, responsible for implementing the society's prejudices, so to speak, by choice" (Hubbard, 1988, p. 232).

"……science takes brains, and that's where we've got it all over them. Science rules the world, Jon. Hitch your wagon to science" (from The Cunning Man, R. Davies, 1994 , p. 45)

The development of new reproductive technologies has been described as "simultaneously liberating and eugenic" (Rapp, 2000, p. 2). They are liberating in that they can provide a means by which some parents may conceive or give birth to the child they desire, and eugenic because they predominantly achieve this end through the selective termination of those foetuses judged to be physically or genetically undesirable. While both vantage points claim the 'moral high ground' as the most socially responsible, by either preventing pain and suffering for, or unconditionally accepting, a child with a disability, it is clear that the practice of prenatal screening uncomfortably straddles a gaping ethical divide.

Many argue that the technologically mediated goal of preventing disability is persistent and defensible, a worthy pursuit that in no way passes judgement on existing people with disabilities and this view seems widely shared in society (Parker, Forbes, & Findlay, 2002; Savulescu, 2001). However, the means by which it is achieved, and the entrenched social prejudices it instils are strongly contested from within the disability rights perspective. When the nature of such 'prevention' comes in the form of terminating an existing fetus or embryo, the disability 'prevention' that flows from prenatal screening takes on a radically different form – a clearly more extreme form of 'prevention' than a measles inoculation or folic acid supplement for example. Rather than preventing disability per se, the consequence of prenatal screening could more accurately be described as the prevention of the live birth of a child with a disability.

If people with disabilities interpret the aims and outcomes of this technology as further stigmatising them, as they seem to (Newell, 2003; Parens & Asch, 1999; Shakespeare, 1998), any humane reflection on its use must include the very voices

Naomi Sunderland et al. (eds.), Towards Humane Technologies: Biotechnology, New Media and Ethics, 113–131.

of those who feel marginalised by it, a point reiterated by Lindemann- Nelson who notes:

"According people the respect they are due is a matter of general importance, and we ought be especially scrupulous when people who have endured a history of negligence and abuse claim that they are yet again being demeaned" (Lindemann - Nelson, 2003, p. 3).

The factors that shape this ethical divide between the cultural imperialism of technology as irrefutably good and the moral unease of prejudicial, discriminatory and stigmatising practices directed against people with disabilities are the focus of this chapter. So, while embracing the 'technological imperative' may hold great promise for the relief of pain and suffering, improved quality of life, and enhanced well being for many, the challenge remains; 'how do we support the humane use of such technologies, in ways that do not further re-enforce discriminatory and stigmatizing practices ?' Our capacity to confront and address this question will shape the ethical landscape we create and the ethical legacy we leave.

The currently favoured biomedical response to such ethical tensions in medical care is to focus on obtaining a patient's informed consent as an indicator of ethical probity and there is widespread support for the ethical credentials of informed consent protocols in the medical discourse. But while informed consent protocols are widely adopted, low levels of informed consent are commonly observed across a broad range of medical interventions, including prenatal screening. Although women appear to actively participate in this practice, research reveals that most cannot articulate the purpose or potential outcomes of testing, indicating low levels of informed consent that undermines any claim to ethical integrity (Bernhardt et al., 1998; Markens, Browner, & Press, 1999; Press & Browner, 1997; Stapleton, Kirkham, & Thomas, 2002; Williams, Alderson, & Farsides, 2002a). In relation to prenatal and genetic screening, the decision to participate is influenced by many things such as the routine nature of screening, overt counselling, the predefining of certain outcomes as unacceptable, the lack of access to alternative care and inadequate education prior to consenting, and these factors converge to create a situation in which choice may be deeply constrained. Therefore, reliance on simple interpretations of individual 'choice' to participate is proving to be a poor basis for establishing ethical probity, and there is a need to further explore and appreciate the underlying factors that shape the ethical dimensions of this practice.

This chapter will explore the ethical tensions that emerge from these opposing interpretations of technology use, as potentially both dehumanising and salvific, and examine the deeply embedded social, personal and institutional considerations that shape genetic and prenatal testing. While the applications and inventions of biotechnology have been described as a 'new' revolution, they are arguably influenced by some very old, historically significant and deeply entrenched prejudices and practices. Collective (mis)understandings of new technologies have colonised our expectations of the controls we can impose on the human condition, shifted boundaries in determining what an 'acceptable' life is, and ultimately influenced the judgements we make of what constitutes a suitable body for our children to be born into. Against a cultural backdrop in which the value of individual choice is paramount, in which an unborn child has no recognised legal (or moral?) status and consumer demands permeate our expectations of clinical care, consideration of how and why such tests are developed, offered and accepted demands further exploration.

SOCIAL CONSIDERATIONS

Enhancing Reproductive Freedom: The Social Construction of "Choice"

Once reserved for only 'high risk' pregnancies, prenatal screening[ii] has now become a routine feature of antenatal care in most first world countries, with up to 90% of women receiving some form of testing during pregnancy (Boyd, Chamberlain, & Hicks, 1998; Clayton, 1999; Dormandy, Hooper, Michie, & Marteau, 2002). Regarded as a means of increasing reproductive choices through enhanced knowledge and more informed decision making (Baird, 1999; Cuckle, 2001), prenatal screening is seen by many parents and doctors alike as unproblematic, even 'responsible pregnant behaviour'(Lippman, 1991, p. 28).

Recent studies showing elective termination rates of 85%-98% for foetuses prenatally diagnosed with some type of physical, chromosomal or genetic anomaly reflect a strong trend towards the abortion of affected foetuses, flagging termination as the dominant response to such unwelcome news (Carothers et al., 1999; Egan & Borgida, 2002; Erikson, 2001; Ford, 1999; Garne, 2001; Santhalahti, 1999). As termination rates are high and the procedure is voluntary, it has been inferred that there is 'no conflict in the practice' (Cuckle 2001, p. 85). Conversely, these same statistics are interpreted as evidence of narrowing options that highlight a lack of perceived alternatives in responding to a diagnosis predefined as undesirable, as Regina Kenen notes:

'Once a pregnant woman is presented with a category of conditions pre-defined as a defect, the social context narrows her perceived options' (Kenen, 1999, para 9).

Others take the position that if we have the means to gain such genetic information about our children, we have a 'parental and moral obligation' to acquire it (Clarkeburn, 2000, p. 1; Green 1997; Savulescu, 2001). However, once such knowledge is known, it cannot be unknown, creating an expectation of action. Clarkeburn clarifies this obligation:

'The duty to acquire information about the genetic constitution of the foetus *must* be connected with the duty to act upon that information .... In the case of untreatable and non-preventable genetic defects, the only possible benefiting action is termination'(Clarkeburn, 2000, p. 2 italics added).

Broader (and perhaps more pertinent) questions of who determines the grey areas of 'non-preventable', or at what level the 'untreatable' becomes 'unacceptable' often remain curiously unarticulated. These considerations are further complicated by the considerable lack of agreement within the medical literature on the definition of what constitutes a disability. For example, obesity, low intelligence, criminality, predisposition to cancer or lack of athleticism (Rothblatt, 1997), are all traits with a genetic component that could be considered debilitating by some, but may not be considered reasonable grounds on which to terminate a foetus or reject an embryo. The problem of applying such malleable definitions of disability further obscures the criteria for use of this technology, particularly in a biomedical and social context in which individual choice is seen as valuable, and medical consumerism drives much of the agenda. The normative outcome of termination may effectively create a situation that limits, rather than enhances, reproductive choice for some.

For many couples, 'choice' inevitably equates to the expedient acceptance of the most common route, a pathway that is clearly paved for them long before they realize they are on it (Lippman, 1999). So, although the terms 'choice', 'consent', 'informed consent' and 'autonomy' are liberally scattered through the medical discourse, in reality, a person's ability to exercise truly autonomous choice is restricted by multiple social, institutional and personal constraints as individuals embedded in time, culture history, relationships and prior understandings. Many of these 'choices' are made under non – negotiable constraints, situated in entrenched institutional cultures and practices of which the person themselves may not even be aware. The question of choice then becomes more accurately one of 'choices made available'. However, the danger that arises in adhering to narrowly defined concepts of agency and choice with their limited orientation is that an overall ethical legitimisation of the established health care routines has occurred while deeper concerns about inherent injustices and failures within existing and unquestioned institutional structures have remained largely unchallenged (Sherwin, 1992, 2001). For the reasons alluded to above, the premise that 'consent' or 'choice' validates ethical engagement has been widely refuted by social researchers as the numerous constraints that act to impede 'choice' in this context are rarely accommodated or even acknowledged in clinical practice.

Routines, Risks and Re-assurances

Further factors obscuring the choice to embrace prenatal[i] screening are the routine nature of screening and the pervasive discourse of 'risk' surrounding pregnancy which creates a need to be re-assured (Lippman, 1991). The reassuringly routine way in which screening is presented implicitly assumes participation and, therefore, obscures the voluntary nature of the proposed test. Consequently, it is harder to opt out of such an accepted and expected routine without appearing 'difficult' or 'uncooperative', a situation that further erodes the validity of any consent given, and calls into question the notion of choice when no alternative is offered (Lippman, 1999; Markens et al., 1999; Press & Browner, 1995, 1997; Santhalahti, Hemminki, Latikka, & Ryynanen, 1998; Williams et al., 2002a). The inclusion of newer tests such as Nuchal Translucency[iii] scanning under the (unquestioned) rubric of an older established antenatal routine (eg, ultrasound), has further blurred the exercising of choice, as the starkly different clinical aims of each test may not be clearly differentiated by some women.

The need for reassurance coupled with the 'desire to know' features strongly in explanations women give of why they participate in prenatal screening programmes (Lippman, 1991; Press & Browner, 1993, 1997). When this perceived need for reassurance is coupled with the widely expressed belief that knowledge is benign and not harmful of itself, a powerful motivation to engage in screening is formed (Anderson, 1999; Press & Browner, 1993; Santhalahti, 1999), However, the belief that *all* knowledge is of value is questioned by Kenen, who remarks, 'if a "gift" of knowledge offers no cure, is it valuable?' (Kenen, 1999, p. 1545). A pertinent example of questionable knowledge is the reporting of ultrasound soft markers[iv]. While clinical reporting of 'soft markers' is commonplace, their relevance is complicated by their often transient appearance. In most cases when a soft marker is identified, a healthy, normal child is born. A comprehensive UK study reported that 92% of foetuses identified as potentially abnormal using soft markers were later confirmed to be normal at birth (Boyd et al., 1998). Boyd's study reported a 12 fold increase in false positives when relying on 'soft markers' with a corresponding increase in diagnosis of actual malformations of only 4% (ibid., p. 1579). Even in this extensive six year study of over 33, 000 pregnancies, only '55% of *(the 725)* malformed foetuses and infants were correctly identified prenatally' (ibid., p. 1578). Considering the high false positive rates and the potential to place healthy foetuses at risk by unnecessary exposure to amniocentesis or unwarranted termination[v], the reporting of soft markers, which are by definition ambiguous and transient, remains controversial, though some practitioners feel it is 'unrealistic and unethical not to report anomalies that may leap to the eye' (Fitzgerald, 1999, cited by Getz & Kirkengen, 2003, p. 2046). This conviction that every 'anomaly' must be disclosed, regardless of its predictive power or veracity, means that the separate clinical aims of determining obstetric risk, revealing structural abnormalities in the foetus, and assessing the risk of chromosomal defects (associated with soft markers), are not selectively pursued in practice. Therefore, the separate ethical considerations surrounding each aim are not articulated, often leading to the disclosure of unsolicited information. While false positives are generally presented as 'good mistakes', the ongoing psychological distress women experience when dealing with such 'false positives' remains an under acknowledged and under researched feature of our commitment to this imperfect technology.[vi]

Closer consideration reveals that of the multitude of potential abnormalities that exist, prenatal screening can identify a relatively small proportion and even then, conditions for which treatment can be offered are relatively few (Baird, 1999; Press & Browner, 1997). Indeed, according to Baird (1999, p. 9) most neonatal disabilities result from 'low birth weight, prematurity, viral or bacterial infection, birth trauma or accident', none of which can be predicted by prenatal screening. Therefore, even a welcome test result in reality offers limited guarantees. In light of these considerations, the importance of providing 'reassurance' to the 94–98%[vii] of women who receive a normal test result would seem to be greatly exaggerated.

If we accept that any reassurance gleaned is of a fairly limited kind, why do women perceive a need to be tested? Many social factors have been identified in the 'construction' of this 'need'. The pervasive language of risk, defect and negative outcomes surrounding pregnancy goes some way to explaining the deeply rooted need to be reassured (Anderson, 1999; Lippman, 1991). Even use of the word 'risk', which means 'the possibility of suffering, harm or loss'[viii] , may convey a strong message of impending danger. Thus the contextual and narrative features of the information given immediately create anxiety and uncertainty. When the negative language of risk is contrasted against the positive expectations of improved prognosis, promises of better medical management and maximised outcomes for baby, accepting the screening on offer appears an obvious route to follow (Lippman, 1991). Secondly, the automatic assignation of the unwelcome 'high risk' label to mothers over thirty five[ix] almost ensures compliance with 'risk lowering behaviours', such as engaging in screening, as membership of this category creates a concern that needs to be addressed (Lippman, 1991). Finally, while clinicians define risk as a population-based number, Getz (2003, p. 2052) points out that 'to the individual pregnant woman, the population base is one, and a risk of one in one hundred means she can be the one'. This personalisation of risk creates a heightened perception of a looming crisis, contributing to the anxiety identified as a motivating factor in compliance with the testing regimes on offer.

Conflicting representations of what constitutes a 'risk' lends further confusion to this dynamic, for example, the risk for a woman over thirty five years old of having a child with Down's syndrome is 1 in 350 (Downs Syndrome Association of Queensland) and is presented by practitioners as 'high risk'. On the other hand, the risk of miscarriage while undergoing amniocentesis to diagnose this condition is 1 in 200 (Alfirevic, 2003), clearly a numerically higher risk, but one that is presented as a 'low risk' procedure. For women trying to grapple with these concepts, malleable interpretations of 'risk', often coupled with poor scientific literacy, may present a barrier to genuine understanding. So, although 90–95 % of affected children are born to mothers with no previous history and most children with Down's Syndrome are born to mothers under thirty five that are considered 'low risk', the prevailing belief seems to be that pregnancy, one of life's ordinary experiences, is a time of heightened risk. Therefore, one of the consequences of accepting this risk discourse is that pregnancy which is a normal, 'well' process has become a 'diagnosis', leaving it open to technological intervention and inspection. Through embracing technology, pregnancy has been simultaneously demystified by opening a window on the previously hidden world of the developing fetus (Rapp, 2000, p.29), then re-mystified by placing it under the exclusive and specialised realm of technological surveillance, open to interpretation only by 'experts'.

Desirability, Disability, Culpability and Responsibility

Interpretations of what level of risk is acceptable, and what measures a person would take to avoid this outcome, depend largely on how undesirable the particular outcome is perceived. Therefore, perceptions of what life with a disability entails, lie at the heart of the decision to embrace prenatal screening. Some women on receiving a diagnosis of Downs Syndrome in their child felt relieved that this was a condition compatible with life, in which their child could lead a relatively meaningful , though supported life (Bridle, 2004). Others on receiving a diagnosis of cleft palate, perhaps a more minor disability, and one of the few 'fixable' conditions detected with no intellectual impairment, immediately elect termination[x]. Clearly differing interpretations of desirability influence such decisions.

The assumption that life is worse than non-existence for the severely disabled is disputed as a socially constructed interpretation, entrenched by the portrayal of disability as an unrelenting tragedy (Clapton, 2003) [xi]. Furthermore, the language of 'coping' and 'burden'[xii] that permeates the discourse surrounding pregnancy and disability are based on a negative presumption of what it means to be disabled, imbuing a message of unrelenting hardship (Clapton, 2004; Reinders, 2000). This message may be further entrenched by the visible and experienced lack of support in the community, confirming the difficulties of living with disability (Ferguson, Gartner, & Lipsky, 2000; Williams et al., 2002a). Given that much of the pain, anxiety and stress experienced by the disabled and their carers directly results from lack of social and governmental support, not from the disability per se (Ferguson et al., 2000; Parens & Asch, 1999), many people with disabilities and their carers feel it is unreasonable to judge quality of life based on a social norm in which they are under-resourced and under supported. As disability rights advocate Tom Shakespeare[xiii] (2001) notes:

> While I support a woman's right to choose, I regret situations where a pregnancy is terminated because of inaccurate or prejudiced information about what it is like to be disabled (Shakespeare, 2001, cited in the Australian Law Reform Commission (ALRC), 2001, p. 38).

While subtle messages about 'desirability' are conveyed through institutional practices, presumptions and protocols, the nature of patient/doctor or patient/counsellor interaction in prenatal screening is persistently idealised as being 'non-directive' and 'non-coercive' (Anderson, 1999). However, the potential to exert influence in this relationship is widely recognised by health professionals who acknowledge the difficulties in maintaining a neutral, non-coercive approach. Some examples offered by Williams' et al (2002b) follow:

> If you offer it in such a way that it may not be such a negative thing to have a baby with Down's, then it may not be....you may get someone else who through past experiences may offer it in a different way, so we have got an incredible amount of power in that relationship. (ibid., p. 746, Midwife).

This sentiment is further re-enforced,

> But then, sometimes, I have had the feeling that people, I wouldn't say are pressurised, but that they are not necessarily given a realistic idea of what the outlook will be for their unborn child, that too black a picture may be painted, and maybe by people who don't actually know themselves. (ibid., p. 746, Paediatrician)

These comments elucidate the impossibility of dispensing unbiased, value-neutral advice in a non-directive way. For many, the mere offer of a test will be viewed as a tacit recommendation (Press & Browner, 1997; Williams et al., 2002a), while the presentation of a diagnosis predefined as adverse, framed in the encumbered language of risk and undesirability, effectively limits choice. While genetic counsellors may be well placed to provide more appropriate education and advice to women, in terms of routine prenatal screening, advice is predominantly given by non genetics practitioners[xiv] who, according to Williams, Alderson and Farsides (2002b) often have little practical experience or knowledge of genetics or disability and hence rely on problem centred textbook accounts, supplementing their educator role with the dispensing of leaflets[xv] (Williams, Alderson, & Farsides, 2002b). The poor levels of understanding that flow from this mode of patient education have been identified as a significant factor in the low levels of informed consent observed (Braddock et al., 1999; Godolphin, Towle, & McKendry, 2001).

Given that most women identify themselves as advocates for their child's health, they accept responsibility for ensuring their child's best interests (Lippman, 1999). Although there may be different expectations of mothering in different social contexts, a consistent theme is that women, more so than men, are responsible for their children's well being (Lindemann-Nelson, 2001, p.136; Rapp, 2000), a responsibility that now begins even before conception with the taking of folic acid supplements. Hand in hand with this sense of responsibility, comes a sense of culpability if a decision not to co-operate in screening regimes leads to a 'negative outcome', that is, an undetected disabled child (Marteau & Drake, 1995 , p. 1128). Consequently a very real awareness of a social pressure to participate in prenatal screening programmes is felt by women. Clearly, the exercising of autonomy in this circumstance incorporates more than simple, rational 'choice' but is constrained by the deeply related and embedded context in which such interdependent 'choices' are made. Ruth Hubbard captures this sense of culpability in saying:

> If a test is available and a pregnant woman doesn't use it, or completes the pregnancy although she has been told that her child will have a disability, the child's disability is no longer an act of fate. She is now responsible; it has become her fault. (Hubbard, 1988, p. 232).

A further institutional consideration shaping technology use and directing choice is the legal framework under which prenatal screening occurs. While there may be nothing legally contentious about a woman giving her informed consent to have her fetus screened prenatally, should she choose to undergo a termination in response to what those tests reveal, her actions, and those of her doctors' often fall under the Criminal Code. While termination of pregnancy is unlawful in most states of Australia, lawful 'therapeutic' termination can be readily procured on the grounds of threats to the physical or psychological health of the mother[xvi]. The presumption that a disability in a fetus is automatically grounds for psychological distress in the mother is firmly entrenched in clinical practice and in law; however, the nature or severity of any fetal anomaly rarely demands justification. There is no evidentiary requirement to substantiate any claim of psychological distress, and as noted previously, termination of a 20 week fetus with a cleft palate is legally defensible on the grounds that it creates emotional anguish in the mother. The boundaries of this pre-condition were tested in a recent Australian case in which a 32 week fetus with 'suspected' achondroplasia (dwarfism) was terminated because the mother was reportedly suicidal. The ensuing public outcry resulted in a police investigation which concluded, as the child was stillborn from the lethal injection it was given in utero, the termination was lawful (de Crispegny & Savulescu, 2004)[xvii]. So, while the nature, severity, or even presence of a disability in a fetus is not specifically recognised as grounds for legal termination, through focusing on the mental distress such disabilities may create in the mother, the law puts women in the unenviable situation where they must build a legal case for termination, based on their own mental fragility. The compounding affront of having conceived a substandard child, followed by the necessity to build a legal case for termination based on ones own mental incapacity inhumanely places mothers at the epicentre of this series of accumulating 'failures' while denying them access to the necessary support to fully consider the ongoing psychological, health and ethical issues at stake.

Further confusion emerges from the laws governing discrimination, wrongful birth and termination, which present conflicting messages about disability. In a recent Australian example, two wrongful birth claims were rejected, on the grounds that there was no substantive evidence that life with a disability was worse than the alternative of non existence[xviii], illustrating the paradoxical nature of laws that reject wrongful birth on the grounds that life with disability cannot be proven to be worse than non- existence, while supporting the terminations of foetuses with severe, minor or no disabilities. A common theme of legal frameworks adopted in many Western countries is that the fetus has no legal status. The trigger for recognising personal rights in law is birth; if a fetus is never born it has no rights and cannot be discriminated against, hence, the prevailing rights are those of the mother. Recent law reform in Australia relating to intentional harm of an unborn child adds a further layer of complexity and contradiction. Under this type of legislation[xix], if an assault on a woman prevents the live birth of a child she is carrying, the perpetrator may receive life imprisonment for the intentional harm of an unborn child. While these laws governing termination are legally consistent in holding the fetus as a 'non- person' and preserving the individual legal rights of the mother, there are real moral inconsistencies in recognising on one hand that the killing of an unborn child constitutes harm, while on the other, condoning the termination of a 32 week old fetus that had substantial chance of life, because the child's mother was experiencing psychological distress. While the current placement of termination under the criminal code reflects the wider social view of termination as undesirable, it appears that terminations can be accessed with relative ease. Therefore, rather than controlling or reducing the numbers of terminations, the current legal framework under which terminations are carried out merely focuses the emphasis on cementing a defensive legal justification; while silencing and sidelining discussion of the moral, human and public health consequences of embracing prenatal screening technologies. Often, the secrecy that flows in such a defensive legal climate, coupled with privatisation and individuation, inhibits open reflection as many of these decisions become deliberately concealed from the public discourse.

When such deeply entrenched social, personal and institutional constraints discussed above are not acknowledged, the language of *choice* can become 'illusory and mocking' for some (Bridle, 2004). Realistically, for many the choice to undergo prenatal screening is the last in a long line of essentially 'non-choices', socially entrenched and predetermined to funnel parents down a particular normative pathway, laid out long before they even conceived their child. Therefore reliance on 'choice' as a measure of ethical integrity is deeply inadequate as it fails to incorporate the many unspoken constraints that underlie any such choice. The powerful culmination of the law, the cultural imperialism of science, and the master narrative of individualism merge to shape the political landscape in which prenatal screening is practiced. A humane response calls for the reshaping of the political space by contesting these dominant discourses.

ETHICAL CONSIDERATIONS: THE ETHICS OF CONSENT AND AUTONOMY

Given the complexities discussed above, it may be helpful to consider how ethical approaches in clinical care are contributing to shaping more reflective and humane technology use. The predominant response within contemporary bioethics to such tension has been to focus on patient autonomy as the measure of ethical probity. If a person chooses a particular pathway, providing they are competent, were not coerced, and have all the relevant information then such a path is generally deemed ethically unproblematic. When Beauchamp and Childress (2001) put forward their now dominant 'principalist' account of ethics, they arguably never intended that autonomy should become the principal principle, however, deference to autonomy is now a dominant feature of the biomedical discourse (Sandman, 2004).

Autonomy has gained precedence for several reasons. Firstly, it re-enforces culturally embedded notions of individual rights, as Charles Taylor notes, it feeds into our "inherited legacy of political thinking in which the notion of rights plays a central part in the justification of political structures and action" (Taylor, 1985). Secondly, as a philosophical principle that can be readily institutionalised into a procedure, the garnering of informed consent neatly fulfils the administrative demands of health care delivery en masse. Thirdly, as a culturally accepted means of placing potentially contentious issues in the hands of individuals, collective accountability can be abrogated, and finally, autonomy sits comfortably beside the minimal legal requirement of information disclosure for informed consent, thus provides a 'ready made' defensive legal position for practitioners. So, the focus on autonomy accommodates many existing, unquestioned and arguably inequitable frameworks while upholding the view of human beings as being detached, rational and atomistic entities, individually exercising their autonomous agency.

However, there are problems with 'consent' as the benchmark of ethicality. Historically, consent has been treated as something of a formality within medical practice (Wear, 1998) and Sherwin (1992) argues that the current attempts to incorporate and layer such shallow ethical understandings onto existing practice is simply a device by which the medical profession can 'demonstrate its serious interest in moral matters to encourage the public to maintain its trust in the physician's judgement' (ibid., p. 86). Thus the analysis of 'ethics' within the narrow parameters of consent and autonomy, entrenched in existing hierarchal medical structures further masks unacceptable and entrenched inequalities while paying superficial respect to moral concerns.

The Autonomous Self

Reasons for this ethical failure may lie in part with the implicit assumptions embedded in the concepts of autonomy, consent and informed consent. Initially flowing from a modern, liberal perspective, the 'self' is seen as a rational, self-directed, un-encumbered, impartial being who is free and able to act independently with no relational or physical constraints (Sherwin, 1992; Taylor, 1985). While other philosophers temper this view with the caveat that particular circumstances and relational constraints must always be considered (Beauchamp & Childress, 2001; Mitchell, Kerridge, & Lovatt, 1996), and fuller accounts of 'relational autonomy' are becoming more widely debated (Mackenzie & Stoljar, 2000), the practicalities and time pressures faced in busy clinics frequently impinge on the capacity to incorporate consideration of these thicker contours of individual lives, entrenching simplistic, perhaps even formulaic approaches to ethics.

Further obscuring the debate surrounding 'consent', 'informed consent' and 'autonomy' as ethical benchmarks, are the pliable definitions and often inter-changeable use of these three distinctly separate concepts. While 'consent' may be defined as an active agreement to embark on a proposed course of treatment, increasingly 'consent' is claimed to be *implied* by participation alone (Cuckle, 1995, 2001). Informed consent incorporates the added dimension of education and the transfer of knowledge prior to the bestowing of consent. 'Informed consent' then is given from a position of understanding after careful reflection and deliberation of the relevant facts. However, who decides what knowledge is to be revealed, to whom and for whose benefit, is largely left unexplored. Again, the power in this exchange rests with the holder of the knowledge (usually the physician) who filters what information is worthy of imparting. While the term 'informed' suggests the mere possession of information, from an ethical perspective, it is not information alone, but rather *reflective understanding*, that creates insight into the critical implications of such knowledge required for autonomous decision making. Hence ethically robust intervention is critically linked to patient education. Therefore, from an ethical perspective, there is a stark difference between fulfilling the minimal legal obligation of dispensing 'information' and meeting the fuller ethical requirement of ensuring 'understanding' of the critical implications of any intervention.

The assumption that equipping a person with all relevant information will produce ethically secure outcomes contains three implicit messages. Firstly, it presumes that good information coupled with good reason will produce a morally acceptable result. Secondly, it presumes that the person deciding is cognisant of what 'relevant' knowledge they need, therefore aware if significant information is missing. Finally, this position presumes that there is such a thing as objective and unbiased facts, apparently unaware that the mere existence of such tests, based on the view of disability as wholly unacceptable, belies any claim of neutrality of information.

A further assumption implicit in the principle of autonomy is that the person themselves, and not the community to which they belong, are the final judge of the acceptability of the decision made. Relationships with others also need to be supportive both psychologically and socially, and there must be no consequence of abandonment or retaliation if a choice is made that others oppose (Oshana, 2001). As previously flagged, the issue of support is recognised as crucial to the ability to exercise autonomy yet when it comes to support for disabled members of society, evidence of rejection, under-funding, under-resourcing and negative stereotypes abound (Bridle 2001; Caplan, 1999; Lippman, 1991; Reinders, 2000). This obvious lack of support further impacts on a woman's capacity to act purely autonomously as she will be dependent on others for support if she should she 'choose' to give birth to a child with a disability.

Adopting notions of choice and autonomy further presumes that real access to a variety of options actually exists and can be readily accessed otherwise choice is not an intentional act but rather acquiescence to the 'least worst' option. Frequently genuine access to a range of options is obscured by the routine incorporation of screening into existing antenatal protocols which undermines voluntariness. Often the prenatal diagnosis of conditions for which there is no cure places women in the untenable position of being complicit in 'choosing' their child's death. Even knowing the child had no chance of survival, some women report ongoing psychological distress after terminating a fetus, partially because of the active role they are required to play in their child's demise[xx]. In the case of prenatal screening at least, the capacity to treat has not kept pace with the capacity to diagnose. Explanations are often incomplete, and the selective information deemed worthy of pursuing is driven by what existing scientific methods can measure, rather than what parents may want to know. Therefore, the choices on offer in response to prenatal screening are rarely based on practical cures, other that the offer of 'therapeutic' termination, meaning that in reality, the options are limited. When collective scrutiny and accountability is handed over to the private realm of individual preference (via autonomy and choice), the technological agenda becomes an essentially self regulating body, closed off to public inspection. Under these conditions, which Margaret Urban Walker describes as 'privatisation' (Urban - Walker, 1998, p.172), the subtly coercive dimensions of these practices can become hidden from public view, as avenues of critique become closed and the 'scientific-technological juggernaut'[xxi] rolls on unchallenged.

Towards a More Humane Ethical Orientation in Clinical Practice

As previously noted, the dominant approach in biomedical ethics currently focuses on principalism which calls upon us to make ethical judgements based on consideration of four ethical principles of autonomy, justice, non-maleficence and beneficence (Beauchamp & Childress, 2001). While the practical application of ethical theories like principalism work best when all the parties have the opportunity and ability to contribute, such a state of affairs rarely prevails in a medical context because some perspectives, usually 'experts', are privileged above others, usually patients. According to Tong (2002, p. 420), 'the resulting situation where some people do all the talking, while others do all the listening impedes the emergence of the full moral truth'.

This leaves us with the question of what type of ethics can accommodate a technology centred culture, thick with diversity and complexity, where the moral topography is shifting. Morality exists in practice not principles, and it is seldom the principle that causes ethical tension. In terms of prenatal screening, perhaps both sides of the ethical divide would agree 'in principle' that life ought to be preserved. Any divergence arises from differing interpretations of when life begins or what quality of life demands preservation. Therefore, it is unhelpful to think of ethics as a collection of competing principles or overarching formulaic rules to be rigidly applied (usually by expert outsiders) in resolving dilemmas. Ethics must be explored as a way of relating and 'being'. Therefore, a practical, applied and embedded approach to enhancing human relationships and understandings is called for, rather than a detached, intellectual, individualistic pursuit.

Margaret Urban-Walker's (1998) expressive–collaborative model goes some way to providing such a framework as it acknowledges the importance of individual accounts in an ongoing, dynamic process of dialogue, collaboration and negotiation progressing towards a more balanced, ethically coherent response to the complexities of prenatal screening. As a narrative approach, it enables the rich context and histories of each situation to emerge and provides a platform from which previously silent voices can speak, where all accounts are heard. Given that ethical decisions are made over time, while the patient is undergoing changes in their emotional, physical and knowledge status (Press & Browner, 1995, para 8), the need to keep moral, reflective spaces open within institutional frameworks is essential in achieving meaningful dialogue (Urban - Walker, 1993, p. 38). As moral truths are rarely delivered in 'one fell swoop, but pieced together, perspective by perspective'(Tong, 2002, p. 423) a framework which allows decisions to be revised and refined must be employed, re-iterating the point that ethical decision-making is 'not a solution to a puzzle, but rather an outcome of a negotiation' (Urban- Walker, 1993, p. 36), a negotiation that is embedded in a time, a context and a history.

In considering the ethical consequences of embracing prenatal screening technologies we are called upon to ask ourselves, what kind of future such interventions will create? Of all our possible imagined futures, which is best? In this sense ethics is unique because it calls us to take an appraisive stance about whether a particular outcome is right or wrong, good or bad, supporting or demeaning, humane or de-humanising (Isaacs & Massey, 1994). Hence inter-dependence and interconnectedness (not individualism) must be regarded as both the starting point and the goal of any ethical conversation. As Eric Cassell notes, "the individual is whole only in a world of others"(Cassell, 1991, p.26). We therefore need to embrace an ethical framework that acknowledges this fundamental human understanding.

SUMMARY

We do ourselves an injury, as individuals and as a society, if we let fear of difference tempt us to decide "who should and should not inhabit the world" because it is hubris to pretend that we have the knowledge and foresight to make such judgments well. (Hubbard, 1988, p. 234).

Clearly technology has brought significant gains to our understandings of the human condition, making substantial contributions to human wellness and wellbeing. Notable examples include the control of infectious diseases, mass immunisation, reduced infant mortality and longer life expectancy. However, these gains have been accompanied by an increasing disposition to regard the body as an object, with science being the culturally iconic lens through which we interpret bodily "malfunctions". The gaze of technology, with its focus on the 'body object' as a malfunctioning machine, has seductively invited us to redefine our understandings of what a suitable body is, and through the recruitment of prenatal and genetic screening has altered our expectations about what constitutes a suitable body to be born into. But bodies are not merely objects to be manipulated or machines to be tuned; they are the site of our human 'being' in the world. As such, any purely mechanistic view of disease or illness that fails to incorporate the essential humanistic understandings, of the deeply connected, embedded and relational realities that define the human condition will fail in its capacity to guide 'humane' technology use.

The various factors discussed throughout this chapter as shaping prenatal and genetic screening each have a thread of connectivity winding through. This common thread is the modernist view of ontology which presents the human self as individualistic, detached, atomistic and rational, who realises 'the good life' through the pursuit of rational thought, knowledge and reason. Under this account, any reflection on the embodied, embedded human realities that funnel which 'choices' are made available remains conspicuously excluded from the wider conversation, while the social construction of 'choice' flowing from this orientation creates a conviction that personal autonomy, self directed and rationally considered, legitimates any action, but as Leon Kass notes "not all human dignity consists of reason or freedom"(Kass, 2002, p. 17).

Therefore a critical pre-requisite in crafting a humane response to the many and complex ethical issues raised by prenatal screening technologies must be the rejection of the individualistic and detached self portrayed by modern moral philosophy. Once we start to accept that we are not essentially disconnected individuals but rather a deeply related, embedded, embodied, and historically and culturally situated human beings, we can more accurately respond to the intricately layered nature of our 'being'. Armed with the understanding that interdependence and interconnectedness, rather than individualism lie at the core of forging any ethical understanding, we will surely be in better position to develop more compassionate and humane approaches to technology use.

NOTES

[ii] These figures refer predominantly to ultrasound and more recently Nuchal Translucency screening, both of which are prenatal screening (not diagnostic) tools to detect potential structural, genetic or chromosomal anomalies. Nuchal translucency is defined as the 'maximum thickness of subcutaneous translucency between the skin and soft tissues overlying the cervical spine' (Nicolaides, 1998, cited by Getz & Kirkengen, 2003, p. 2049) Increases in this measurement, taken between 11 to 13 weeks of pregnancy are often linked to chromosomal or genetic anomalies in the fetus. Most of the developmental work on this technique has been focused on the detection of Downs syndrome (See also Snijders, Noble, Souka, & Nicolaides, 1998).

[iii] See endnote 1.

[iv] Soft markers are defined as 'Structural changes detected at ultrasound scan which may be transient and in themselves have little or no pathological significance, but are thought to be more commonly found in foetuses with congenital abnormalities, particularly karyotypic abnormalities' (Bricker et al., 2000 cited by Getz & Kirkengen, 2003, p. 2046) Examples include echogenic foci of the heart, echogenic bowel, choroid plexus cysts and shortening of limbs.

[v] Two foetuses were terminated without amnio after disclosure of soft markers was made. Subsequent genetic testing after termination revealed no chromosomal abnormality. (Boyd et al., 1998).

[vi] Several participants interviewed in a current research project, who had experienced false positive from Nuchal translucency, describe the associated trauma and depression. They noted ongoing and niggling doubts about the health of their children, which they attributed directly to this suspicious result. From "Enhancing Ethical Practice in Prenatal Screening – Facilitating Women's Ethical Choices" (Milligan, 2007)

[vii] See Baird, 1999; Boyd et al., 1998; Carothers et al., 1999; Ford, 1999

[viii] Microsoft, 1999

[ix] Anecdotally, there seems to be a downward pressure on the age at which a woman is categorised 'high risk'. Classified by RANZCOG as above 38 y.o., accepted clinical practice now defines women over 35y.o. as high risk. One woman interviewed in current research project "Enhancing Ethical Practice in Prenatal Screening – Facilitating Women's Ethical Choices", was told her 'advanced maternal age' of 31 y.o was a risk factor that should compel her to undergo screening (Milligan, 2007)

[x] A case reported during research interviews with Queensland Health Professionals. In this instance, the termination was not carried out by Qld Health, but the couple were referred to a private provider to access termination (Enhancing Ethical Practice in Prenatal Screening – Facilitating Women's Ethical Choices" , Milligan, 2007)

[xi] Clarifications of what constitutes 'severe' are a further source of contention.

[xii] The word 'burden' is defined as 'something emotionally difficult to bear, a responsibility or duty, to weigh down or oppress' (Microsoft, 1999).

[xiii] Tom Shakespeare lives with the medical condition achondroplasia or dwarfism.

[xiv] There is little research on this issue of who provides information for prenatal genetic screening. Research emerging from "Enhancing Ethical Practice in Prenatal Screening – Facilitating Women's Ethical Choices" (Milligan, 2007), suggests that a very small proportion of women actually see a genetic counsellor and screening is so routine and unquestioned; the normative pathway compels women to screen first and ask questions later. Part of this dynamic is that, while genetic counsellors may be the best placed professional to dispense this advice, genetic information is typically explained by another practitioner who may have inadequate genetic knowledge themselves. A further issue diminishing informed consent is the lack of continuity of care and resulting blurred accountability of which practitioner is responsible for patient education.

[xv]Leaflets alone have been found to be quite ineffective in patient education if not supplemented by the opportunity to engage in dialogue with a practitioner (Braddock, Edwards, Hasenberg, Laidley, & Levinson, 1999; Godolphin, Towle, & McKendry, 2001)

[xvi] See http://www.childrenbychoice.org.au/nwww/auslawsum.htm for summary of various Australian State Laws relating to termination.

[xvii]Melbourne obstetrician Lachlan de Crispegny has been identified as the person who performed this termination. He fails to disclose his involvement in this particular article, but discusses his pivotal role in the book "Forever on Trial" (see article by Melinda Tankard Reist in Online opinion http://www.onlineopinion.com.au/view.asp?article=2560 accessed Nov 2006)

[xviii] Harriton v Stephens. 2006 HCA 15; 9th May 2006; Waller v James; Waller v Hoolahan. 2006 High Court of Australia 16; 9th May 2006

[xix]Section 313 of the Criminal Code of Queensland.

[xx] While there is some research into the psychological consequences of termination, there is very little specific research into the psychological impact on women of terminating a planned and wanted pregnancy on the grounds of an unplanned and unwanted condition or disease. This research is needed to develop appropriate support structures for women experiencing 'therapeutic' termination. 'High levels' of psychological distress after termination of foetuses with prenatally diagnosed conditions, for up to 12 months were noted by V. Davies, Gledhill, McFadyen, Whitlow, & Economides, 2005. There are many internet support groups on which women share their post termination stories such as, Born too soon http://www.borntoosoon.freeservers.com/photo6.html,, the Heartbreaking Choice support group http://www.aheartbreakingchoice.com/ and other perinatal bereavement support groups such as http://www.pbso.ca/ to name a few. In the article "The hardest thing I have ever done", Emma Loach describes her 'momentous decision …to terminate…. and the ferment of relief, guilt and grief' that continues to follow. http://www.guardian.co.uk/weekend/story/0,3605,966128,00.html.

[xxi] A phrase coined by Pullman, Bethune, & Duke, 2005.

REFERENCES

Alfirevic, Z. (2003). *Early amniocentesis versus transabdominal chorionic villus sampling for prenatal diagnosis*. The Cochrane Library.

Anderson, G. (1999). Non-directiveness in prenatal genetics: Patients read between the lines. *Nursing Ethics*, *6*(2), 126–136.

Baird, P. (1999). Prenatal screening and the reduction of birth defects in populations. *Community Genetics*, *2*(1).

Beauchamp, T., & Childress, J. (2001). *Principles of biomedical ethics* (5th ed.). New York: Oxford University Press.

Bernhardt, B., Geller, G., Doksum, T., Larson, S., Roter, D., & Holtzman, N. (1998). Prenatal genetic testing: Content of discussions between obstetric providers and pregnant women. *Obstetrics & Gynaecology*, *91*(5), 648–655.

Boyd, P. A., Chamberlain, P., & Hicks, N. R. (1998). 6-year Experience of prenatal diagnosis in an unselected population in Oxford. *The Lancet*, *352*(9140), 1577–1581.

Braddock, C. H., Edwards, K. A., Hasenberg, N. M., Laidley, T. L., & Levinson, W. (1999). Informed decision making in outpatient practice. Time to get back to basics. *JAMA*, *282*(24), 2313–2320.

Bricker, L., Garcia, J., Henderson, J., Mugford, M., Neilson, J., & Roberts, T., et al. (2000). Ultrasound screening in pregnancy: A systematic review of the clinical effectiveness, cost effectiveness and women's views. *Health Technology Assessments*, *4*(16).

Bridle, L. (2001, August). Confronting the distortions: Mothers of children with Down Syndrome and prenatal testing. *Down Syndrome Association of Queensland*, 19–23.

Bridle, L. (2004). *Stories of choice: Mothers of children with Down Syndrome and the ethics of prenatal screening*. Unpublished PhD, University of Queensland, Brisbane, Australia.

Caplan, A. (1999). What is immoral about eugenics? *BMJ*, *319*, 1284.

Carothers, A. D., Boyd, E., Lowther, G., Ellis, P. M., Couzin, D. A., & Faed, M. J. W., et al. (1999). Trends in prenatal diagnosis of Down syndrome and other autosomal trisomies in Scotland 1990 to 1994, with associated cytogenic and epidemiological findings. *Genetic Epidemiology*, *16*, 179–190.

Cassell, E. (1991). Recognising suffering. *Hastings Centre Report*, May–June, 24–31.

Clapton, J. (2003). Tragedy and catastrophe: contentious discourses of ethics and disability. *Journal of Intellectual Disability Research*, *47*, 540–547.

Clapton, J. (2004, October 8). *Burden, bioethics and disability*. Queensland University of Technology, Applied Ethics Seminar.

Clarkeburn, H. (2000). Parental duties and untreatable genetic conditions. *Journal of Medical Ethics*, *26*(5), 400.

Clayton, E. W. (1999). What should be the role of public health in newborn screening and prenatal diagnosis? *American Journal of Preventive Medicine*, *16*(2), 111–115.

Cuckle, H. (1995). Cost effectiveness of antenatal screening for Cystic Fibrosis. *BMJ*, *311*, 1460.

Cuckle, H. (2001). Extending antenatal screening in the UK to include common monogenic disorders. *Community Genetics*, *4*(2).

Davies, R. (1994). *The cunning man*. London: Penguin Books.

Davies, V., Gledhill, J., McFadyen, A., Whitlow, B., & Economides, D. (2005). Psychological outcome in women undergoing termination of pregnancy for ultrasound-detected fetal anomaly in the first and second trimesters: a pilot study. *Ultrasound Review of Obstetrics and Gynaecology*, *25*(4), 389.

de Crispegny, L. J., & Savulescu, J. (2004). Abortion: Time to clarify Australia's confusing laws. *Medical Journal of Australia*, *181*(4), 201–203.

Dormandy, E., Hooper, R., Michie, S., & Marteau, T. (2002). Informed choice to undergo prenatal screening: A comparison of two hospitals conducting testing either as part of a routine visit or requiring a separate visit. *Journal of Medical Screening*, *9*, 109–114.

Egan, J., & Borgida, A. (2002). Screening for Down syndrome. *Obstetrics & Gynaecology*, *27*(2), 22–30.

Erikson, S. L. (2001). Post-diagnostic abortion in Germany: Reproduction gone awry, again? *Social Science & Medicine*.

Ferguson, P., Gartner, A., & Lipsky, D. (2000). The experience of disability in families: A synthesis of research and narratives. In E. Parens & A. Asch (Eds.), *Prenatal Testing and Disability* (pp. 72–94). Washington, DC: Georgetown University Press.

Fitzgerald, T. (1999). Antenatal Screening: women are being given incomplete information. *British Medical Journal*, *318*, 805.

Ford, N. (1999, September 17). Ethical aspects of prenatal screening and diagnosis. In N. Ford (Ed.), *Conference Proceedings "Scientific, Medical, Ethical and Legal Aspects of Prenatal Screening and Diagnosis"*. Caroline Chisholm Centre for Health Ethics.

Garne, E. (2001). Different policies on prenatal ultrasound screening programmes and induced abortions explain regional variations in infant mortality with congenital malformations. *Fetal Diagnosis and Therapy*, *16*(3).

Godolphin, W., Towle, A., & McKendry, R. (2001). Challenges in family practice related to informed and shared decision - making: A survey of preceptors of medical students. *JAMC*, *165*(4), 434–438.

Green, R. M. (1997). Parental autonomy and the obligation not to harm ones child genetically. *Journal of Law, Medicine and Ethics*, *25*, 8.

Hubbard, R. (1988). *Eugenics: New tools, old ideas*. New York: Haworth Press.

Isaacs, P., & Massey, D. (1994). *Mapping the applied ethics agenda*. Paper presented at the third annual meeting of the Association for Practical and Professional Ethics, Cleveland, Ohio.

Kass, L. R. (2002). *Life, liberty and the defence of dignity. The challenge for bioethics*. San Francisco: Encounter Books.

Kenen, R. (1999). Disability rights and prenatal genetic testing. *Network News*, *24*(2), 3.

Lindemann - Nelson, J. (2003). *Hippocrates' maze. Ethical explorations of the medical labyrinth*. New York: Rowman & Littlefield Publishers Inc.

Lindemann-Nelson, H. (2001). *Damaged identities: Narrative repair*. Cornell University Press.

Lippman, A. (1991). Prenatal genetic testing and screening: Constructing needs and reinforcing inequities. *American Journal of Law and Medicine*, *27*(1&2), 15.

Lippman, A. (1999). Choice as a risk to women's health. *Health, Risk and Society*, *1*(3), 281–291.

Mackenzie, C., & Stoljar, N. (2000). *Relational autonomy. Feminist perspectives on autonomy, agency and the social self*. Oxford: Oxford University Press.

Markens, S., Browner, C., & Press, N. (1999). 'Because of the risks': How US pregnant women account for refusing prenatal screening. *Social Science & Medicine, 49*(3), 359–369.

Marteau, T. M., & Drake, H. (1995). Attributions for disability: The influence of genetic screening. *Social Science & Medicine, 40*(8), 1127–1132.

Microsoft. (1999). Encarta Encyclopaedia.

Milligan, E. (2007). *Enhancing ethical practice in prenatal screening: Facilitating women's ethical choices.* Unpublished PhD-Ongoing research, Queensland University of Technology, Brisbane.

Mitchell, K., Kerridge, I., & Lovatt, T. (1996). *Bioethics and clinical ethics for health care professionals* (2nd ed.). Social Science Press Australia.

Newell, C. (2003). Embracing life: Ethical challenges in disability and biotechnology. *Interaction, 16*(2), 25–33.

Nicolaides, K. H. (1998). Having the test gives parents options. *British Medical Journal, 317*(749).

Oshana, M. A. L. (2001). The autonomy bogeyman. *Journal of Value Inquiry, 35*(2), 209.

Parens, E., & Asch, A. (1999). The disability rights critique of prenatal genetic testing. *The Hastings Center Report, 29*(5), S1.

Parker, M., Forbes, K. L., & Findlay, I. (2002). Eugenics or empowered choice? Community issues arising from prenatal testing. *Australian and New Zealand Journal of Obstetrics and Gynaecology, 42*(1), 10–14.

Press, N., & Browner, C. (1993). 'Collective fictions': Similarities in reasons for accepting maternal alpha-fetoprotein screening among women of diverse ethnic social class backgrounds. *Fetal Diagnosis and Therapy, 8*(Suppl.), 97–106.

Press, N., & Browner, C. (1995). Risk, autonomy and responsibility: Informed consent for prenatal testing. *The Hastings Center Report, 25*(3).

Press, N., & Browner, C. (1997). Why women say yes to prenatal diagnosis. *Social Science & Medicine, 45*(7), 979–989.

Pullman, D., Bethune, C., & Duke, P. (2005). Narrative means to humanistic ends. *Teaching and Learning in Medicine, 17*(3), 279–284.

Rapp, R. (2000). Testing the woman, testing the fetus. *The social impact of amniocentesis in America.* New York: Routledge.

Reinders, H. (2000). *The future of the disabled in liberal society: An ethical analysis.* Notre Dame, IN: University of Notre Dame Press.

Rothblatt, M. (1997). *Unzipped genes: Taking charge of baby making in the new millennium.* Philadelphia: Temple University Press.

Sandman, L. (2004). On the autonomy turf. Assessing the value of autonomy to patients. *Medicine, Health Care, and Philosophy, 7*, 261–268.

Santhalahti, P. (1999). Participation in prenatal screening and intentions concerning selective termination in Finnish maternity care. *Fetal Diagnosis and Therapy, 14*(2), 71–79.

Santhalahti, P., Hemminki, E., Latikka, A., & Ryynanen, M. (1998). Women's decision making in prenatal screening. *Social Science & Medicine, 46*(8), 1067–1076.

Savulescu, J. (2001). Procreative beneficence: Why we should select the best children. *Bioethics [H.W. Wilson - SSA], 15*(5/6), 413.

Shakespeare, T. (1998). Choices and rights: Eugenics, genetics and disability equality. *Disability and Society, 13*(5), 665.

Shakespeare, T. (2001). *The danger of disability prejudice.* Retrieved September 1, 2001, from www.genecrc.org.site/hi/hi3z.html

Sherwin, S. (1992). *No longer patient. Feminist ethics and health care.* Philadelphia: Temple University Press.

Sherwin, S. (2001). Diagnosis difference: The moral authority of medicine. *Hypatia, 16*(3), 172.

Snijders, R., Noble, P., Souka, A., & Nicolaides, K. H. (1998). UK multi-centre project on assessment of risk of trisomy 21 by maternal age and foetal nuchal translucency thickness at 10–14 weeks gestation. *The Lancet, 352*(9125), 343–346.

Stapleton, H., Kirkham, M., & Thomas, G. (2002). Qualitative study of evidence based leaflets in maternity care. *British Medical Journal, 324*(7338), 639–643.

Taylor, C. (1985). *Philosophy and the human sciences: Philosophical papers 2.* Cambridge: Cambridge University Press.

Tong, R. (2002). Teaching bioethics in the new millennium: Holding theories accountable to actual practices and real people. *Journal of Medicine and Philosophy, 27*(4), 417–432.

Urban - Walker, M. (1993). Keeping moral space open: New images of ethics consulting. *The Hastings Centre Report*, *23*(2), 33–40.

Urban - Walker, M. (1998). *Moral understandings: A feminist study in ethics*. London: Routledge.

Wear, S. (1998). *Informed consent: Patient autonomy and clinician beneficence within health care* (2nd ed.). Washington, DC: Georgetown University Press.

Williams, C., Alderson, P., & Farsides, B. (2002a). Too many choices? Hospital and community staff reflect on the future of prenatal screening. *Social Science & Medicine [H.W. Wilson - SSA]*, *55*(5), 743.

Williams, C., Alderson, P., & Farsides, B. (2002b). What constitutes 'balanced information in the practitioners' portrayal of Down's syndrome? *Midwifery*, *18*(3), 230–237

Eleanor Milligan
School Here
University Here

INGER LASSEN

9. CONSTRUING 'HOPE' IN GENE MODIFICATION DISCOURSE

A critical study of implicit mental processes

INTRODUCTION

Over the past 20 years, many Europeans have come to see food biotechnology as a threat to humanity and nature in spite of reassurances by scientists that the new technology offers many promises and ought to be hailed as a panacea for a wide range of world problems. The debate has been pursued in the media, by political elites and what Bauer and Gaskell (2002: 227) have referred to as 'attentive publics' in their book about controversial biotechnology issues. In the debate that, more often than not, focuses on possible risks associated with food biotechnology, many concerns and fears are voiced. But since the counterpart of fear is hope, the two sentiments live happily together in an intertextual web of sensing processes that may take on various forms. These preliminary thoughts resonate with Kant's questions (cited in Rose 2007: 257), 'What can I know? What must I do? What may I hope? These questions demand ethically justified answers from a number of participants, including citizens, authorities, professionals, bio-tech companies and not least the media.

This chapter reports on a study of language resources used by stakeholders in a contestation of claims and demonstrates how linguistics may be useful in critically analyzing the discourses focusing on issues involving ethical dimensions. In eleven interviews conducted by a Danish freelance journalist (Christiansen 2002), 'attentive publics' and GMO experts express their different hopes and fears. The interviewees represent varying attitudes, but nevertheless seem to agree on the common goal of making the world a better place in spite of diverging opinions about how this goal may be reached. Taking its starting point in Halliday's (1994) fundamental speech roles of giving and demanding information, my theoretical approach foregrounds the Appraisal framework, which is a framework offering resources for evaluating attitudes (Martin and White 2005; White 2001). These resources unfold in an intertextual (Fairclough 2003) web that interacts with semantics in interesting ways. Having analysed various realizations of 'hope', I suggest that in some of its realizations the mental process 'hope' may be categorized as a desiderative process with a meaning close to 'want', and in other meanings as a cognitive process that comes close to 'think'. Besides, in some of its realizations 'hope' interacts with appraisal resources in interesting ways and thus expands the meaning potential of the texts. The implication of such meaning

Naomi Sunderland et al. (eds.), Towards Humane Technologies: Biotechnology, New Media and Ethics, 133–150.

expansion is that in the interviews I have analyzed, genetically modified organisms (GMOs) are constructed as a humane technology through the many implicit and explicit resources offered by language, - resources which function as ideological vehicles for attitudes that invade our sometimes unconscious minds, thus blindfolding us to some of the processes affecting GMO-related issues. Certainly, the general lack of transparency within the field of GMO, combined with limited access to knowledge about possible risks, will have detrimental effects on our trust in authorities and producers.

THEORETICAL APPROACH

For an analysis of expressions of hope, Halliday's (1994) fundamental speech roles seem to be a natural point of departure. If clauses are exchanges of meaning, as Halliday suggests, speakers are involved in interactive events based on the speech roles of i) giving or demanding information and ii) giving or demanding goods and services (Halliday: 2004; 107, revised by Matthiessen). In the GMO-interviews I have analyzed, information is demanded by the interviewer and given by the interviewee mainly through statements and questions as illustrated in table 1:

Table 1. Giving or demanding goods-&-services or information

Role in exchange	Commodity exchanged	
	(a) goods-&-services	(b) information
(i) giving	'offer' [Proposal]	'statement' [Proposition]
(ii) demanding	'command' [Proposal]	'question' [Proposition]

(Halliday (revised by Matthiessen) 2004: 107)

Interestingly, a question may at times be realized as a command, and a statement may occasionally have the force of an offer. Such blending of speech functions is well established in pragmatics and has been referred to as interpersonal Grammatical Metaphor (GM) within Systemic Functional Linguistics (Halliday, 1994: Ch. 10)

In the system of speech functions, the notion of modality as 'intermediate degrees' between the outer poles of negative or positive statements plays a central role in that it reflects the speaker's degree of commitment to a given utterance and thus positions the speaker dialogically, thereby constructing subjectivity in the texts (Iedema, Feez & White (1994) (White 2001). As a result, the speakers' subjective values are expressed on a cline between a 'yes' and a 'no' to GM technology. Halliday (in Matthiessen 2004) makes the important distinction of propositions (statement and questions) and proposals (offers and commands). Propositions and Proposals may be modalized through modal verbs or adjuncts to realize i) degrees of probability, ii) degrees of usuality, iii) degrees of obligation, iv) degrees of inclination and what Iedema, Feez & White (1994) have referred to as v) potentiality. Further, a useful approach to studying the subtleties of value expressions is offered by the Appraisal framework referred to earlier and developed by scholars from the Sydney school, notably Martin 1997; Martin and Rose 2003; Martin and White 2005; Iedema, Feez and White 1994; Rothery and Stenglin 1997; Coffin 1997; Droga and Humphrey (2002). Moreover, Halliday's theory of Grammatical Metaphor (GM) (1994, 1998, 2004) constitutes a valuable tool for the analysis of meaning making resources used to express hope, - a point I shall demonstrate in what follows.

METHOD AND DATA

To bring to the fore some of these points, I analysed eleven interviews undertaken by a Danish freelance journalist on behalf of a research information centre, Biotekcentre. In the interviews, scientists, politicians and a highranking represent-ative from a Danish consumer organization presented their views on genetic engineering and biotechnological issues. The interviews were later compiled in a book sponsored by Biotekcentre – a Centre established in a joint effort by bio-chemical companies including Aventis, DuPont, Monsanto, Plant Science Sweden, and Syngenta. The freelance journalist was given the following task: 'Present a picture of the current attitudes to GMO and biotechnology in Denmark through interviews with approximately equal numbers of skeptics and supporters' (2002: 8). The interview design was made by the freelance journalist himself, and intertextuality seems to have played an important role in the technique he used.. The interview design was based on a 'devil's advocate' approach, turning replies by GMO-sceptics into questions to GMO-proponents and vice versa. This may be illustrated in the following sample dialogue:

Interviewer prompting GMO-opponent. 'Wouldn't that (**organic production**) increase the shortage of food? (2000: 19)

Response from GMO-opponent. 'In many places in the third world, the land is not utilized to the full, and there a conscious bet on **organic farming** would improve the harvest' (2000: 19)

Interviewer prompting GMO-proponent. 'I guess there has to be room for **organic farming** also?' (2000: 50)

Comment from GMO-proponent. 'Of course - if the consumers want **it**, and if they will pay for **it**' [organic farming: author's comment] (2000: 50)

Interviewer prompting GMO-opponent. 'Won't the issue be solved by the **market forces**? In our part of the world, consumers can afford still more sophisticated products while in the third world most people are satisfied with more healthy food products at the cheapest possible price ...' (2000: 66)

Comment from GMO-opponent. 'When decisions are made by **the market**, it is definitely **not always to the benefit of the general public**.' (2000: 66)

Interviewer prompting GMO-proponent. 'Representatives of the officially declared opponents against GMOs say that **the poor in the third world won't get their share** of the economic advantages of GMOs....' (2000: 90).

These prompts and comments serve to illustrate that the dialogue unfolded around selected topics. As for organic farming, the GMO-proponents preferred the markets to decide, while the GMO-opponents preferred organic farming rather than genetically modified farm products. The question-answer sequence moreover indicates diverging positions as to who might possibly benefit from genetic modification of food.

Seemingly, to lend a voice to diverging positions, a journalist was given the task of making interviews with approximately equal numbers of proponents and opponents. However, although at a first glance there seemed to be a balanced weighting of conflicting views, a close reading suggested that on a cline of acceptance, six interviewees were positive towards GMO, two were moderately positive, and two were predominantly skeptical and expressed greater fear than the rest of the interviewees. This is borne out already through the headlines, which form a web of intertextuality throughout the eleven interviews.

Intertextuality of Headlines

The following interview headlines, sub-categorized according to stance, offer a preview of some of the attitudes expressed:

In favour of GMO
- Organic farmers should recognize and use the GM technology
- Compare GM crops, organic and conventional agriculture – and then pass your sentence
- Better food safety, flora, fauna and environment are good arguments for GM technology
- Danish skepticism towards GMOs makes biology students opt out of plant cultivation
- When we allow doubt to stop us, we let stupidity reign
- We can afford to worry, we are not starving – that is typically Danish

Moderately skeptical towards the use of GMO
- We are both for and against GMO, but above all we are cautious
- It is crucial that GMOs are of genuine value – and not just to the industries and patentees

Skeptical towards the use of GMO
- When it comes to plants, Denmark should remain a GMO free area
- Doubts that GM crops will benefit consumers

The attitudes foreshadowed in the ten headlines above may be placed on a cline of acceptance as shown in figure 1 below:

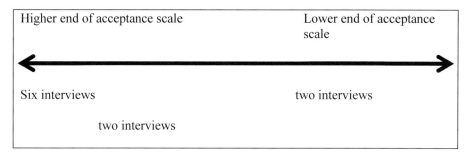

Figure 1: Ten interviews placed on a cline of acceptance

In addition to orienting the reader towards the contents of each interview, the headlines also served the purpose of ensuring textual coherence in the book of interviews as each headline offered a preview of the essence of the attitudes expressed.

Intertextuality of KeyTopics

A closer look at intertextuality made it possible to identify an interrelated web of conflicting stances, which focused on repeated key topics central to the bio-technology debate. A list of topics selected from the interviews included: harvest yields, plant diseases, safety, ethical dilemmas, the environment, biodiversity, profitability, organic farming, the third world, research and development funding, the public debate, the media, patents, consumers and competition from the US, most of which hold elements of risk as well as ethical dilemmas.

The variation of attitudes towards GMO technology in the eleven interviews was particularly noticeable when looking at different ways of expressing hope and fear. Hope was expressed more frequently in interviews favoring GMO technology. In these interviews hope was directly associable with benefits that GMO technology might possibly offer. If fear was expressed in GMO-friendly articles, the focus was on detrimental effects caused by the opponents' lack of acceptance. In the interviews positioned at the lower end of an acceptance cline, 'hope' was not expressed as frequently. Moreover, if referred to at all, hope was associated – not so much with GMO technology as with methods or legislation that could minimize possible risks. In interviews with participants who were mainly skeptical, fear played a much more significant role and expressions of fear mainly related to GMO technology risks, multinationals and patent rights, thus displaying distrust in bodies responsible for governance.

ANALYSIS AND DISCUSSION

Before we proceed into an actual analysis of expressions of hope, let us look at what Halliday would refer to as the most congruent way of expressing hope, viz. by using the mental verb 'to hope'. Mental verbs are divided into the following four types of SENSING: *perceptive, cognitive, desiderative* and *emotive*, on the basis of the way they differ in terms of grammatical features and metaphorical potential. A fairly clear distinction is made between, on the one hand, *cognitive* and *desiderative* mental verbs and, on the other, *emotive* and *perceptive* verbs (1999: 138). An important distinction is that cognitive and desiderative verbs can create worlds of ideas, as distinguished by Halliday and Matthiessen (ibid.). However, while cognitive verbs project ideas about information that may not be valid, desiderative verbs project proposals that have not materialized but whose materialization depends on desire (ibid: 140). When an interviewee makes a statement such as:

> *Example 1.* 'I hope that with time, GM technology will be recognized and used by environmentalists' (2000: 10)

The projecting clause 'I hope' projects (a desire that something be the case) through the Mental verb 'hope', which - according to the distinction above – is a cognitive verb because it projects ideas about information that may or may not be valid and whose materialization does not depend on desire.

Throughout the eleven interviews there were only three examples of interviewees using the word 'hope'. Of these, two expressed 'hope' through the sensing verb 'hope' while the third example was expressed as a nominalization of the verb. This is shown below in:

> *Example 2.* At least I have hope (2000: 60) in which the Sensing verb is rendered as noun..

However, as we shall see, there are other ways of hoping for a better world, and in what follows I shall offer an overview of the realization options used in the interviews and then discuss some of the realizations on the basis of the Appraisal framework illustrated in table 2:

Table 2. Realization of hope in ten interviews

HOPES (in direct quotes)			
ENGAGEMENT			
ATTRIBUTED UTTERANCES 3 examples	ATTITUDE		
	AFFECT	JUDGMENT	APPRECIATION
	7 examples	7 examples	5 examples
MODALITY (67 examples in total)			
10 examples	9 examples	43 examples	5 examples
DISCLAIMERS			
3 examples	0 examples	0 examples	0 examples
PROCLAIMERS			
3 examples	0 examples	0 examples	0 examples

As shown in table 2, very few realizations of hope were attributed to speakers other than the interviewee. At a first glance this would indicate a strong commitment on the part of the speaker. However, this impression was counteracted by the frequent use of modality, counting 67 examples in all. The vast majority (43 examples) were modifications of JUDGMENT (a category for appraising people's actions) while fewer examples of modality were found for the categories AFFECT (a category for appraising people's emotions) and APPRECIATION (a category for appraising things). The frequent use of modality might be explained by certain semantic properties of the sensing verb 'to hope', which has been sub-categorized as a desiderative verb with the distinctive feature that it can project ideas and thus create worlds (Halliday in Matthiessen (2004: 210). A closely related sensing verb is that of 'to think', which is sub-categorized as a cognitive verb that can project ideas. There is thus in both desiderative and cognitive mental verbs inherent semantic features that involve prediction and thus a degree of uncertainty, reflected in the speakers' ways of talking about future issues. In what follows, I shall exemplify the interplay of such verbs with a number of Appraisal resources:

Appraising People's Actions: Examples of Modalized Judgment

Typical instances of modified judgmental statements used for creating a world as a better place to live are shown in the following examples. Examples 3-6 show various realizations of *probability* and *capacity*.

> *Example 3. By means of genetic engineering* we <u>can do</u> this [interfere with nature: author's comment] in a better, faster, safer and more applied manner. <u>I am convinced</u> that future generations will make the conclusion that GM technology has contributed greatly to making the world a better place (2000: 13) (Judgment in *italics*; <u>modality underlined</u>).

The propositions are uttered in reply to a question by the interviewer as to whether genetic engineering does not mean interfering with God's own mechanisms. The argument furthered here is that 'genetic engineering' is a positive thing that will help future generations. This is what the interviewee hopes will happen, without mentioning the word hope. By using the modal verb 'can', to modify the verb 'do', the speaker makes a positive Judgment of scientists' capacity for creating a new plant by means of genetic engineering. The judgmental value is reinforced by 'I am convinced', which expresses strong support of the proposition without making a full commitment. Co-text plays an important role in evoking the Judgment of capacity as illustrated by the example. In other words, what we 'can do' will also be done in order to reach the ultimate goal of 'making the world a better place'. By stressing this point, the interviewee implicitly constructs the future world of GMO technology as a positive world.

> *Example 4.* <u>I think</u> that in a few years DFL-Trifolium will be ready with *adapted, environmentally beneficial grass varieties for Danish agriculture and export* (2000: 60)

In example 4, the cognitive verb, I think, expresses the speaker's opinion as to the probability of DFL-Trifolium launching 'adapted, environmentally beneficial grass varieties' in a few years to come. By describing DLF-Trifolium's invention as 'environmentally beneficial' he praises the company's integrity and moral standards, thus creating a world he hopes will materialize.

> *Example 5*. I don't think the media have created the public concern, I think the media *reflect this concern* (2000: 44)

In example 5, 'I think' – a Grammatical Metaphor for 'probably' – is used by the speaker to express his attitude towards the media, which are often blamed by GMO-proponents of serving their own purposes and not those of the general public. The speaker refers to the media in positive terms, indicating that through the verb 'reflect' the media is constructed as an institution of integrity that provides an accurate representation of general public sentiment. A more congruent realization of 'I think' is found in the modal adjunct 'probably', shown in example 6. Here, the speaker voices a hope of 'achieving the same result' as has been achieved with recombinant antibodies for therapeutic use from mammalian cell cultures, which is positive Judgment of using a technology that has been found beneficial and useful for medical purposes. That it can be done 'at almost no cost' contributes to evoking hope that such an invention would be beneficial. However, the value of the hope expressed is reduced from median to low by the modal verb 'could', and the disclaimer 'but that is not allowed' further reduces the force of the proposition.

> *Example 6*. By using GM technology on plants, we could probably *achieve the same result at almost no cost*, but that is not allowed (2000: 56)

Examples 7-16 are examples of judgmental propositions containing modal verbs of obligation. They are presented on a cline from the more congruent to the more metaphorical expressions of evoked hope. In examples 7 and 8, the issue of technological advance is commented on. In example 7 below, the modal verb 'must' expresses a subjective orientation with a high-value degree of obligation. The claim (Toulmin 1995) 'it must be positive towards new things' is backed by the proposition 'it is a liberal society' and the conditional clause 'if it has a tradition' and warranted implicitly by the commonly accepted idea that 'liberal societies have traditions'. The positive aspect of the Judgment passed here is that 'being positive towards new things serves a good purpose and is commendable behavior earning the label of integrity. The speaker thus, implicitly expresses hope that the world he has created, will eventually dawn.

> *Example 7*. It is a liberal society. If it has a tradition, it must *be positive towards new things* – like technological advances (2000: 91).

In example 8, the modal verb 'will have to' expresses objective obligation on a line with 'is required to'. The speaker, presupposing that GMO technology is beneficial to the environment and 'entirely safe', thus resorts to hoping that Greenpeace will lose the battle. He thus implicitly passes negative Judgment on Greenpeace, an impression that is furthermore strengthened by the prejudice contained in the wording 'Greenpeace and the like', which makes the assumption that Greenpeace and similar organizations that engage in environmental activism cannot be trusted. However, for this interpretation it is necessary to rely on knowledge about the world.

Example 8. When the population understands that the technology benefits the environment and sustainability – and at the same time is entirely safe – Greenpeace and the like <u>will have to</u> *pack their bags. They will lose the battle* (2000: 91)

The modal verb 'should' in example 9 is used by the speaker to blame the media, represented here by the public service TV channels, for not doing what they are supposed to do, viz. to inform the general public in an unbiased way. In the source text, the media are described by the speaker as having contributed to an unqualified debate, which has scared the general public. According to the speaker, they should instead allocate more time to the GM issue. The media are thus deemed incompetent, which significantly lowers their social esteem through negative Judgment of incapacity, unlike the attitude towards the media expressed in example 5 above.

Example 9. They [the media: author's comment] <u>should</u>, as a run-up to the political and ethical discussion, *make a series of programmes which for months follow the work in the researchers' laboratories* […] (2000: 28)

Examples 10-13 deal with four different realizations of the mental verb 'need'. The four realizations share the feature of assessing the situation as one of *not having* or *being without* something that the speaker finds important, useful and necessary. The realizations – in their different forms – are thus full of hope for something that is judged positive. However, in the context and due to the absence of the phonomena needed, the attitudes expressed must be categorized as negative.

In example 10, for instance, the speaker hopes for the passing of rules and control to optimize safety in terms of GMOs.

Example 10. There is no such thing as complete safety. But <u>we need</u> *carefully worked-out rules for both research and production, and a strong public approval and control apparatus, which can use international rules as its point of departure* (2000: 65)

A similar hope is expressed in example 11, however through nominalization of the process 'need' which is rendered instead as a Grammatical Metaphor through the noun 'necessity'. In an unpacked version, the clause might be rendered subjectively as 'we need to follow the same procedure of approval'. But in its realization as GM, it becomes possible for the speaker to make an objective claim and at the same time insert the qualifier 'utter' in front of 'necessity', thus strengthening its force.

Example 11. It is <u>an utter necessity</u> *that we all follow the same procedure of approval* (2000: 30).

Example 12 is similar to 11 in that it involves a Grammatical Metaphor 'the use of GMOs necessitates', which might be unpacked as 'If we want to use GMOs, we need everybody – including the consumers – to have the possibility of making an actual choice'. In this example a strongly felt hope is expressed that something be the case.

Example 12. The use of GMOs necessitates that *everybody – including the consumers – have the possibility of making an actual choice* (2000: 66)

In example 13, the sensing verb 'need' is rendered as a noun. This makes it possible for the speaker to present 'the needs of the manufacturers' as given information and an undeniable fact, unlike in an unpacked version such as 'as a representative of the consumers I also have to point out that the manufacturers need organic foodstuffs' where it would be possible to deny the projected clause.

Example 13. As a representative of the consumers I also have to point out the needs of the manufacturers *of organic foodstuffs* (2000: 67).

In 14 and 15 we see examples of alternative representations of the desiderative sensing verb 'need'. We may say that 'a prerequisite' in 14 and 'a precondition' in 15 carry meanings that are semantically close to the process 'need' and its nominalized versions 'a need' and 'a necessity'. Both examples express various forms of obligation. The difference between 'necessity' and 'prerequisite' is mainly one of formality, the latter being higher on a scale of objectivity than the former. Thus rather than saying 'I hope many patents will be taken out because we need them to further research and development of GM technology', the speaker formulates his claim more objectively.

Example 14. Patents are a prerequisite to *further research and development* (2000: 76)

A similar use of a nominalized concept is seen in 15 where 'a precondition' is the semantic choice made out of a number of options that inherit features of strong requirement bordering the positive pole of obligation. When we say that something is 'a precondition' for something, we want to convey the meaning that something is strongly needed before something else can take place. An example of this is seen in 15 where the speaker expresses the implicit hope that certain communication barriers between biotechnology researchers and the general public concerning plant biotechnology will go, in order to ensure a better understanding on the part of the general public. However, compared to example 14, the speaker's claim is weakened through the modality in 'can be seen as' where 'can' expresses probability.

Example 15. A multifaceted and long-term effort in relation to these barriers to communication can be seen as a precondition *for development of a comprehensive, specific and fundamental dialogue between biotechnology researchers and the general public on evaluation and use of biotechnology in relation to plants* (2000: 80).

In 16 the speaker's hope is conveyed implicitly through the proposition 'the decisive element will be'. As we saw in examples 14 and 15, the speaker resorts to an objective variant of obligation at the higher end of the obligation scale. So, instead of saying 'in my opinion the researchers must not make the mistake of saying …' the speaker's attitude is expressed implicitly as negative Judgment, indicating potential communication incapacity on the part of biotechnology researchers.

Example 16. The decisive element will be *that the researchers do not make the mistake of saying that GM technology is 'completely safe' because nothing is* (2000: 83)

In example 17 there is a degree of determination involved in the speaker's claim that public control is desirable to achieve 'a better harvest', which is the goal of the speaker's hope. Together with the modal verb 'will' the co-text contributes significantly to the general orientation of the text and helps position the speaker as someone who hopes for a scenario where GM-seeds are available for distribution, however under some public control.

Example 17. Along with the purchase of industrially produced commodities, public control of the production of seed grain will – also qualitatively – contribute *to a better harvest* (2000: 29)

Examples 18-20 may be categorized as Judgment evoked through potentiality, a category that Halliday (in Matthiessen 2004: 621) has characterized as being 'on the fringe' of the modality system. Examples 18 and 19 have different orientations in that 18 'be able to' is an objective implicit variant of the subjective implicit 'can' in example 19. In 18 the speaker conveys the belief that organic plant growers hold the capacity of reaching their goals if (only) they would use GM technology.

Example 18. Organic plant growers would be able *to reach more of their goals more quickly and better by using GM technology* (2000: 9)

In 19 the speaker again praises GM technology, but now drawing on capacity through the implicit subjective modal verb 'can' to express what 'we' (the scientists) are capable of doing. Like in 18 the adverb 'with GMOs' is a condition for successfully realizing an evoked hope of making better, healthier and cheaper foodstuffs - a goal most observers would find acceptable.

Example 19. With GMOs we can make *better, healthier and cheaper foodstuffs* (2000: 73)

Example 20 is slightly different from example 19 in that in 20, the modal verb 'could' is a borderline case between potentiality and probability. The scenario imagined is a hypothetical one in which the availability of GMOs is presented as a condition for release of their potential, and the modal verb 'could' combined with 'help' instigates hope that something be made the case.

Example 20. GMOs <u>could</u> [...] <u>help</u> *give the third world a Volkswagen, and some people in the industrialized world* <u>would be able to get</u> *a slightly bigger car* (2000: 59)

Other Examples of Judgment

Examples 21-23 are instances of implicit positive Judgment relying on resources other than modality. In 21, a lexical metaphor, 'a far stronger bet', expresses the attitude that organic development is preferable to GM technology when thinking in terms of the third world. A probable interpretation of the clause is that 'if only we optimize organic development, the third world would stand a better chance'. In this interpretation, the speaker implicitly expresses the hope that organic development will be optimized – an interpretation warranted by the commonly held belief that the third world should be given better opportunities.

Example 21. Optimizing organic development is <u>a far stronger bet</u> for the third world (2000: 19-20)

A similar attitude is voiced in example 22. In both examples 21 and 22, it is possible to identify desiderative elements of obligation. In 22 this is done implicitly by appraising those who show caution when dealing with GMO, which in the view of the speaker would be a commendable act, showing positive integrity.

Example 22. Caution with such a serious issue as genetic engineering is <u>a virtue, not an error</u> (2000: 69).

In example 23, hope is expressed through the perceptive sensing verb 'I see', which is in this case directed at a desirable phenomenon 'the usefulness'. The co-text 'if we could offer draught and salt resistant plants or crops, etc.' contributes to the positive orientation of the text in that such properties are viewed as beneficial to society.

Example 23. I said that GMO *must prove itself useful before we let go.* <u>I see the usefulness</u> in many third world *countries if we could offer draught and salt resistant plants or crops with a nutritional composition that fights disease and improves the general health* (2000: 40).

Appraising Emotions: Affect

Affect is a gloss for appraising personal feelings. These have to do with emotional states and may be positive or negative. The speaker in example 24 is favorably disposed towards GMO and in his view too little has been done to compare organic crops with GM crops, which – in his view – may have contributed to a biased GMO debate. The hope he raises in the interview is that once the general public would understand that GMO crops are friendly to the environment, GM technology would meet greater societal acceptance. To convey this point of view, the speaker uses a desiderative sensing verb 'want' in a rather strong proposal signifying intention or plan. At the same time this implies that too little has been done in terms of making scientific experiments, thus evoking at the same time negative

Judgment of incapacity because he implicitly blames science, including in particular organic farmers, for not having researched the area sufficiently.

Example 24. I want *experiments in which organic crops are compared with conventional and GM crops respectively* (2000: 23)

In examples 25 and 26, positive Affect is expressed through the modality 'would be nice', a proposal constituting a mild form of a command. Instead of saying 'give me' the speaker uses the politer form of 'it would be nice to get', which expresses the emotional state of happiness at the thought of what the GM variety would be able to bring about in terms of reducing the frequency of spraying.

Example 25. It would be nice *to get a GM variety with resistance against fungus and mould so I could reduce the frequency of the pesticide treatments from the present eight sprays* (2000: 47)

Example 26 is a variant of example 25. Both are polite forms of a proposal using the command 'give me', but while 25 is an objective proposal, 26 is uttered subjectively by the subject person 'we'. Both examples are metaphorical because imperatives such as 'give me' and 'cure us of our diseases' are realized as declaratives with the value of commands.

Example 26. In wealthy Europe we are not starving, but we would like *to be cured of our diseases!* (2000: 77)

A further example of implicit Affect is found in example 27 where 'I would rather place my bet on' indicates a subjective preference on the part of the speaker. The lexical metaphor 'to place a bet' involves a degree of risk-taking. However, the object of 'placing a bet', viz. 'quality, health and food safety' offers security, and the desirability of this choice is emphasized by the adverbial 'rather', indicating an element of opportunism in the speaker, who by placing his bet on 'quality, health and food safety' is led by his inner emotions and hopes.

Example 27. I would rather place my bet on *quality, health and food safety* (2000: 66)

In example 28 an emotive sensing verb uttered by the interacting subject 'I' evokes positive Affect. In welcoming a bio-ethical discussion, the speaker indicates that he hopes for a discussion that would shed more light on ethical issues relating to genetic engineering.

Example 28. I welcome *such a debate* [a bio-ethical discussion: author's comment] (2000: 61).

In example 29 the interacting subject 'I' pays a tribute to the company DLF-Trifolium through the lexical metaphor 'I raise my hat to DLF-Trifolium'. The metaphor is an indication of respect, which reflects the speaker's admiration of and positive emotions towards the company. The relative clause explains why the speaker finds that DLF-Trifolium has earned his respect: They 'dare admit that they are interested in and bet on GM plants'. The speaker thus admires the company's courage and integrity.

Example 29. I raise my hat to DLF-Trifolium, which, as the only Danish company dares to admit that they are interested in and bet on GMO plants (2000: 34).

Appraising Things: Appreciation

Appreciation is a third category in the Appraisal framework. When a speaker appreciates something, the focus is on things, not on people's behavior. Assessing objects from the point of view of Appreciation may involve emotional impact, composition or significance, including the social value, of a thing. In examples 30-32 we find three examples of Appreciation. In example 30 the modal verb 'would' modalizes 'a clear advantage'. By modalizing the utterance, the reduction of the use of pesticides is made probable and a condition for gaining 'a clear advantage'. The condition inherent in the utterance determines whether the speaker's implicit hope of reducing the use of pesticides and fertilizer will be fulfilled, thus adding social value.

Example 30. It would be a clear advantage to the environment *to reduce the consumption of pesticides* and optimize the delivery of fertilizer (2000: 49)

In example 31 'GMO technology' is appreciated in positive terms as a technology that 'can solve' malnutrition and starvation problems. It is thus looked upon as a technology having great significance and social value.

Example 31. GMO technology can solve *that problem* (the problem of malnutrition and starvation) for us (2000: 73)

Along the same lines, significance and social value are appreciated in example 32, where the speaker ascribes 'a high practical value' to 'medicine based on genetic engineering'. The co-text is important when understanding the implications of the evaluation expressed in the speaker's claim. In the subsequent clause the speaker explains the essence of the 'high practical value' referred to in 'we are able to help weak and sick people and to improve and prolong life' if medicine is based on genetic engineering.

Example 32. Medicine based on genetic engineering has a high practical value compared to the uncertainties involves. We *are able to help weak and sick people and to improve and prolong life* (2000: 18)

Establishing Dialogic Position: Engagement

Disclaimers. The implication of the conditional clause in example 33 'but if we are to fully utilize the potential', is that GMO technology is something desirable with a potential we may hope to be able to utilize. This hope is modified, though, by adding that 'a certain degree of acceptance by consumers and society is necessary'. Thus without the consumers' acceptance, GMO technology cannot be fully utilized. GM technology is thus appreciated as something valuable, but its utilization potential rather than its social value is disclaimed, making the speaker's stance clear.

*Example 33.*But *if we are to fully utilize the potential, a certain degree of acceptance by consumers and society is necessary* (2000: 49)

Proclaimers. When a speaker proclaims, he 'interpolates himself into the text as the explicitly responsible source of the utterance' to borrow wording from White (2001: 4 of 15). In a formulation like in example 34, the speaker puts himself at risk of rejection, and engages in dialogue by opening up the discourse to external voices. By saying 'let us run the risk', the speaker agrees with GMO opponents that there may be risks involved, but the noble cause of 'saving the world from social misery and environmental catastrophy' is a strongly motivating goal that makes potential risk dwindle on a balance of risk and opportunity. 'Let us' is closely related with the Imperative mood, but includes both 'you' and 'me'. It may thus be seen as halfway-house between an Offer and a Command (Halliday in Matthiessen, 2004: 139). 'Let us run the risk' is a suggestion that we all give priority to the possible advantages because of their significance to humankind.

Example 34. Let us run the risk *– and help save the world from social misery and environmental catastrophy* (2000: 71).

A variant of 'proclaiming' is illustrated in example 35:

Example 35. How long will it be before *genetically engineered plants are just as accepted and sought after in Denmark and Europe as GMO medicine is?* [....] It is coming. I have no doubts whatsoever (2000: 89).

In example 35 proclaiming is realized by means of different linguistic resources. The speaker asks how long time it will take before GMO plants are accepted at a level with GMO medicine, and he then adds an afterthought proclaiming what he believes to be a fact in 'it is coming. Like in 34, the speaker confronts the risk of rejection by using a cognitive sensing verb 'I have no doubts', thus opening up the scene to external voices, which may or may not agree. The forcefulness of the claim is motivated by the implicit hope that the world is going to be formed as desired. Research has shown that GMO medicine is widely accepted in Europe, including Denmark, and the rhetorical question suggests that the speaker would find it desirable if genetically engineered plants were accepted to the same extent as GMO medicine.

As the examples show, the debate among participants in the GMO-debate has taken a number of directions and conflicting attitudes have been voiced. These vary a lot and subsume goals such as offering better economic opportunities to companies, protecting the environment, improving governance, fighting diseases, improving food quality and helping the developing countries. In the interviews, these goals were rendered implicitly through modalized propositions and proposals, constructing the production of genetically engineered plants as likely, possible, necessary, useful and recommendable acts of integrity. However, at the same time contesting voices were constructing the same undertaking as risky, unnecessary, obstructing free choice, non-beneficial and unhealthy. The attitudes expressed give rise to a number of ethical concerns. In an article Ebbesen (2006, pp 32-50) draws parallels between bioethics and nanoethics, suggesting that the nano-ethical debate does not have to start from scratch, but can draw lessons from the biotechnology debate. Ebbesen (ibid. p 35), quoting two American ethicists who have published widely on bio-ethics (Beauchamp and Childress 2001), refers to a claim that 'a limited number of basic ethical principles is generally accepted'. These ethical obligations or principles include respect for autonomy and integrity, beneficence (promoting benefits), non-maleficence (not inflicting harm), and justice (Beauchamp and Childress 2001). The underlying principles are that a person's autonomy and integrity should be respected, which means that in the case of genetically modified food, consumers should be able to possibly choose not to buy GMO food if this goes against their conviction. A related issue is whether consumers' integrity should not also be respected when it comes to their attitudes concerning environmental protection. Biotechnologists have claimed that if consumers were educated and better informed about the complexities of GMO, they would not be as skeptical. However, studies have shown that more knowledge and information does not necessarily bring about more positive attitudes as attitudes seem to be closely related to the specific applications of GM-technology (see e.g. Bauer and Gaskell 2002).

When considering the principles behind beneficence and non-maleficence, we invariably must include the dimension of risk in our discussion. Beauchamp and Childress (2001, cited in Ebbesen 2006) see risk as closely related to benefit and harm and they suggest that the two elements be compared in risk-benefit analyses in order to find out whether the benefits of introducing a genetically modified product will outweigh negative consequences. This perception is reflected in some of the dilemmas introduced in this paper, such as the dilemma of making a choice between organic food and GMO-food. GM-technology might for instance offer solutions that would reduce the use of pesticides, but at the same time there is no guarantee that plants will not cross-pollinate if grown in the fields.

The last ethical principle in Beauchamp and Childress's list is *justice*. This is seen in the shared attitude of obligation to help third world countries to more healthy lives and to free the world of hunger and starvation. In the interviews, everybody seems to agree that there ought to be at least elements of justice in the world and that people should all have access to food every day. The principle is an all-pervading feature in the interviews and is frequently used by stakeholders (producers, experts, authorities) to build trust and establish integrity. This, however, is not possible without strong communication skills, and communication is also seen as crucial when it comes to gaining public acceptance. A European report (European Commission (2004), quoted in Ebbesen (2006) expresses concern over public reception of nano-technology in the wake of the controversies over GM-technology. The report recommends two-way communication which takes attitudes expressed by the general public into account. It also recommends that the scientific community improves its communication skills. At this juncture, it seems important to raise a critical voice cautioning against the use of communication skills and a programme of ethics to 'insulate researchers from criticism, and from the detailed examination of the nature and consequences of their activities' (Rose 2007: 256). In Rose's wording (ibid.), 'Bioethics often seems to arise from an alliance perhaps an unhealthy one, between those who want or need an ethical warrant for their commercial or scientific activities'. It is of paramount importance that we do not see ethics reduced to what Ebbesen (2006) has called a means to an instrumental end, which can be expressed as a reduction of ethics to a PR agent for the laboraty'. I agree with Ebbesen on this issue and see the role of Humanities and Social Sciences as one of critically examining and deconstructing fundamental issues relating to GM-technology. This would entail asking questions such as what kind of a society we want and how GM-technology and other new technologies will relate to such a society. The outcome does not have to be a negative one. If we carefully balance benefits and harm, using all the commonly accepted principles of ethics, we may have a chance to define goals we can accept and to steer towards these goals.

CONCLUSION

In this article I have analyzed different realizations of the sensing verb 'hope'. Interestingly, in the eleven interviews that represented my data, I found only three examples of hope used as a verb. This does not mean, however, that the concept of hope was not somehow expressed. In fact it seemed to be an all-pervading feature – alongside with fear. Lexico-grammatically, hope took on a number of different forms spanning from various sub-categories of cognitive, emotive and desiderative sensing verbs, to grammatical metaphor through nominalization. Discourse-semantically, hope was construed through modalized appraisal resources, which included *par excellence* Judgment of integrity, but examples of Affect and Appreciation were also found. More often than not, the appraisal resources were implicit and combined with modal verbs or adjuncts. Appraisal resources were generally used in a positive sense by GMO-proponents describing genetically modified products and the activities undertaken to promote GMO-crops. GMO-opponents, on the other hand, used positive appraisal about attempts to ensure food safety and organic farming. Negative appraisal was used by GMO-proponents about the GMO-opponents' skeptical attitudes and by GMO-opponents about some aspects of genetically engineered food. The analysis of appraisal in the eleven interviews thus elucidates the variation of stance among the interviewees and unravels the ideologies implicitly conveyed through various resources afforded by language.

A relevant question to ask at this point is whether my analysis of the concept of hope has offered new insights into what it takes for discourse to make the world a better place to live. The interviews studied exemplify a dialogue between opponents and proponents, mediated by an interviewer. The two parties to the dialogue have diverging opinions, but they obviously share the same goal, viz. a world free of human suffering. Reaching this goal takes a continuous dialogue of competing attitudes, - a dialogue that has to take into account diverging attitudes towards bio-ethics. In this dialogue the interlocutors will participate in an ongoing process of meaning negotiation with the common aim of finding possible solutions to a problem that hesitates to go away. They hope for different world orders in their attempts at finding solutions to the dire problems in which humans are situated, and only if we can agree on what is good and true will we be able to assess whether 'hope' is sufficient when it comes to creating humane rather than human technologies.

REFERENCES

Bauer, M. W., & Gaskell, G. (Eds.). (2002). *Biotechnology. The making of a global controversy*. London: Cambridge University Press.

Beauchamp, T. L., & Childress, J. F. (2001). *Principles of biomedical ethics* (5th ed.). Oxford: Oxford University Press.

Christiansen, J. L. (2002). *11 attitudes to the crops of the future*.

Coffin. C. (1997). Constructing and giving value to the past: An investigation into secondary school history. In F. Christie & J. R. Martin (Eds.), *Genre and institutions. Social processes in the workplace and school* (pp. 196–230). London and New York: Continuum.

Droga & Humphrey (2002). *Getting started with functional grammar*. Berry: Target Texts.

Ebbesen, M. (2006).What can nanotechnology learn from the ethical and societal implications of biotechnology? In *Forum TTN (Journal of the Institute Technology-Theology-Natural Sciences (TTN)* (Issue 15, pp. 32–50). Germany: Ludwig-Maximilian University of Munich.

European Commission. (2004). *Towards a European strategy for nanotechnology*.

Iedema, R., Feez, S., & White. P. R. R. (1994). *Media literacy, Disadvantaged schools program*. Sydney: NSW Department of School Education.

Halliday, M. A. K. (1998). Things and relations: Regrammaticising experience as technical knowledge. In: J. R. Martin & R. Veel (Eds.), *Reading science: Critical and functional perspectives on discourses of science* (pp. 185–235). London: Routledge.

Halliday, M. A. K. (1994). *An introduction to functional grammar*. London: Arnold.

Halliday, M. A. K. (2004). *An introduction to functional grammar*. London: Arnold. (in Matthiessen, revision).

Martin, J. R. (1997). Analysing genre: Functional parameters. In F. Christie & J. R. Martin (Eds.), *Genre and institutions. Social processes in the workplace and school* (pp. 3–39). London and New York: Continuum.

Martin, J. R., & Rose, D. (2003). *Working with discourse. Meaning beyond the clause*. London: Continuum.

Martin, J. R. & White, P. R. R. (2005). *The language of evaluation. Appraisal in English*. UK: Palgrave.

Rose, N. (2007). *The politics of life itself. Biomedicine, power, and subjectivity in the twenty-first century*. Princeton, NJ and Oxford: Princeton University Press.

Rothery, J., & Stenglin, M. (1997). Entertaining and instructing: Exploring experience through history. In F. Christie & J. R. Martin (Eds.), *Genre and institutions. Social processes in the workplace and school* (pp. 231–263). London and New York: Continuum.

Toulmin, S. (1995). *Uses of argument*. Cambridge: Cambridge University Press.

White, P. (2001). *Appraisal: An overview*. Retrieved February 7, 2007, from www.grammatics.com/appraisal/

Inger Lassen
Department of Language and Culture
Aalborg University

SECTION IV RESPONSES

PETER ISAACS

10. THE TECHNOLOGICAL IMPERATIVE AND THE ETHICAL IMPULSE

Exploring the interface

INTRODUCTION

It is frequently claimed that the rapidly developing inventions and applications being generated in biotechnology are indicative of a major revolution in human history whose impact on humankind will be as significant and enduring as earlier technological revolutions. Often, this reference to previous technological revolutions is proposed as an implicit guarantee or assurance that the new revolution is a continuation of humankind's evolution through the inventiveness and application of technology and that, as with earlier revolutions, the welcoming of such developments will only speed the benefits available to humankind. Thus, the advocates for the biotechnological revolution promise a new age of increased human well-being. The relief of pain and suffering, the overcoming of world hunger, an extended quality of life, the generation of new employment possibilities and a growth in economic prosperity are some of the benefits promised. For both the expert and popular imaginations it is a fundamental given that the bio-technological revolution is ethical in nature and that its overriding goal is the diminution of human pain and suffering and the enhancement of human wellbeing.

One consequence of this view that the biotechnological revolution is an eminently ethical engagement, is that it serves to silence critics or doubters. How could one accept as ethically authentic or rationally committed the person who would doubt the benefits of this new technology, not only in terms of future promise, but also in the face of established achievements? Yet, both the confidence in the certain benefits of the biotechnological revolution and the marginalisation of critics and doubters appear as odd social responses when viewed within the tradition of modern reason. As every student of Descartes well knows, the modern tradition of critical reasoning begins from a position of systematic doubt and contemporary studies in cognitive science, epistemology and the philosophy of science have provided greater insight into the complexities of human knowledge and have shown that notions like those of objective truth, certainty and predictability are both complex and often far from attainable in human affairs.

In this chapter, I wish to suggest that the ethical confidence displayed by both the expert and the popular imaginations is not misplaced within the broad ethical tradition of Western culture. This tradition has either been neutral as regards technological interventions or it has welcomed such interventions as authentic

Naomi Sunderland et al. (eds.), Towards Humane Technologies: Biotechnology,
New Media and Ethics, 153–171.

contributions to human wellbeing. In short, while one might seek to challenge the validity of many of the epistemic claims of biotechnology, that such technology is beneficial and above ethical scrutiny is, within the Western ethical tradition, not a matter of dispute. In the same way that technology can be seen as shaped by its history, so also might the Western ethical tradition be seen as shaped by its history. And what I am suggesting is that, for most of that history, Western conceptions of the good life have favorably accommodated or endorsed the technological impulse. Of course, this not to suggest that such ethical endorsement stands firm and unchallenged today for I wish to argue, also, that more recent conceptions of the good life provide a platform for serious ethical critique of the contemporary technological and biotechnological impulses. While the expert and popular imaginations are still fired by the past conceptions within Western history and culture of the positive interrelationship of technology and the good life (ethics), more recent conceptions of the good life point to an emerging rupture in that relationship.

In this chapter I wish to explore the present biotechnology-ethics interface by drawing from the perspective of history. I wish to suggest that, from its earliest beginnings and until well into the twentieth century, the technological imperative has not been seen as ethically problematical. Indeed, it has been seen as more and more central to the Western conception of the good and civilized life. However, I wish to suggest, further, that the present ethical challenges facing the technological imperative reflect a rupture between the technological imperative and the ethical impulse as a result of post-modernity and the emergence of a post-modern ethical sensibility. Specifically, I wish to provide, first, an historical overview of the technological impulse in shaping the human condition. Secondly, I would like to provide also a brief historical overview of the Western ethical impulse, the Western ethical tradition. Thirdly, I would like to offer some basic reflections on the interface between the historical development of technology and the historical development of Western ethics. Fourthly, and finally, I would like to provide some final reflections on the ethical challenges facing technology and biotechnology from a position of post-modern ethical sensibility. My goal is broadly Socratic. I wish to invite those who see the good life as necessarily being enhanced by the technological impulse to pause and to examine that life, to consider how that life incorporates conceptions of the good that have been historically shaped, and to ponder how emerging conceptions of the good might be more critical of the claimed benefits and beneficence of the new technologies.

THE TECHNOLOGICAL IMPULSE – A HISTORICAL OVERVIEW

An important area of interest in both anthropological and historical studies is the way in which human cultural evolution has been shaped by technology. Pekka Kussi in his 1986 study This World of Man (sic) sees humans as constantly changing living organisms whose evolution is shaped by both biological and cultural factors. Kussi suggests that the history of humankind is a history of a growing and more dominating role of culture in shaping both the human present and the human future, that this culture is primarily shaped by technology, and that the primary goal of such technology has been the enhancement of human life and survival by overcoming the constraints implicit in humankind's nature as biological, embodied and nature-dependent beings. In broad terms one could characterise human evolution from the emergence of homo sapiens as moving through three cultural stages – hunter-gatherer cultures, agrarian cultures and scientific-technological cultures.

Hunter-gatherer cultures

Present humankind is believed to have evolved from homo sapiens, a species which is thought to have appeared in Africa more than 50,000 years ago. These early human communities were hunter-gatherers. Anthropologists have proposed that they probably lived in tribes of about forty members and were sustained by the land through the gathering of wild plants, berries, nuts and fruits, the hunting of birds and animals and the fishing of rivers, lakes and seas. These early communities were deeply aware of the place of nature in shaping the human condition since nature provided them with all the requirements of life – food, clothing, shelter, medicines, and implements. Yet, the overriding and constant goal of hunter-gatherer communities was that of survival and the successful pursuit of this goal required that the hunter-gatherer devote almost all of their collective endeavors to this task. This required that each tribe deal satisfactorily with three continual challenges: the gathering of food, the protection of the tribe from threats (whether from nature or other tribes) and the renewal of the tribe through reproduction. This latter needed to be balanced against the capacity of the surrounding environment to provide for increases in members since hunter-gatherer communities lived in a defined geographical area. Fundamental to this goal of living in ecological harmony with their natural surrounds was the acquisition and transmission of a sophisticated expert knowledge of their land. This expert knowledge involved a richness of detail, for it involved an intimate, particular and encyclopedic knowledge of the land in terms of meeting the needs of the community from year to year and from season to season. Such expertise reflected a collective tribal knowledge and wisdom accumulated over many millennia of living with, and listening to, the voice of the land. While hunter-gatherer communities have largely disappeared, their traditions still survive in the Indigenous peoples of the world today.

Agrarian cultures

It is thought that humans moved from a nomadic, food gathering, and tribal lifestyle to a settled lifestyle focussed around settlements and the cultivation of crops and animals about 10,000 years ago, about 8,000 BCE. Agrarian culture allowed communities to settle by developing two secure methods of acquiring food, the cultivating of plants and the herding of animals. Villages and towns were gradually established and systems of trade and barter developed. Central to the life of these agrarian communities was a need for permanency and this was realized by living on the land rather than off the land. Thus, the new communities redefined the relationship between humankind and the land as they sought to dominate nature and to maximize their exploitation of nature's resources as demanded by the generation of surplus. While the new techniques of farming and agriculture were able to realise often an abundant supply of food, at times such strategies failed and failed catastrophically. Fire, earthquake, drought, flooding and pestilence could suddenly occur and easily destroy. Furthermore, settled communities were ripe for attack from other coveting communities. Thus, the move to agrarian communities also redefined the relationship between communities as potentially hostile and vulnerable. The reality of external aggression emerged as a threatening constant in the vulnerability of settled communities.

The production of surplus food, the settled nature of agrarian life and the opportunity provided to disseminate knowledge and information provided for more complex forms of social organization. Kuusi sees the historical evolution of agrarian cultures as progressing through four major stages of evolving social complexity – c 8,000 – 3,500 BCE, the period of small states; c 3,500 – 600 BCE, the period of early empires; c 600 BCE – 500 CE, the period of classical states; 500 CE – 1,750 CE, the period of European dominance (Kuusi, 1986: 90). Within these stages Kuusi suggests that the second period of 3,500 – 600 BCE is the period which sees the most rapid consolidation of the agrarian lifestyle. This period witnessed the transition in human social organization from settlement or town, through the development of fortified cities or city-states, to that of a unified system of governance which extended across vast geographical areas and included numerous communities and towns. Historians refer to this new social phenomenon, which first appeared in lower Mesopotamia about 3500 BCE, as the emergence of civilizations or empires. Such empires were ' based upon the state with leaders, slaves, oxen, plows, wheels, horses, metals, temples, coins and writing skills' (1986, 107). It was the integration of social, economic and technological forces which underpinned these early agrarian empires and which provided the key features of subsequent empires up to the time of the Industrial Revolution.

Furthermore, the power of these empires required a substantial production of surplus to support the non-agricultural classes, a need which was met through the acquisition of more distant goods either by trade or by forceful acquisition. This latter strategy was the most efficient means for expanding the production of surplus and involved either displacing hunter-gatherer communities from their lands or the conquering and subjugating of other established agricultural communities. In other words, the generation of surplus, (which underpinned the contributions of non-farming classes), when coupled with the aspiration of all classes to a more affluent lifestyle, necessitated a further social dynamic of colonization to generate the additional resources required. Hence, these early empires were characterised by both agricultural success and military prowess.

Since the good life for agrarian societies presupposed sustainability, technological developments that enhanced agricultural success and military success were seen as necessarily contributing to the good life. And once social stability and sustainability were assured, then the contribution of technology to enhancing the production of lifestyle goods and artifacts were seen as enhancing the quality of life.

Scientific-technological cultures

The emergence of a scientific-technological culture is marked by a major change in the means of production. Within agrarian cultures goods were produced primarily through the work of humans, oxen, horses and water. The change introduced with this new culture was firstly that of using machines to replace the traditional forms of labour. James Watt's steam engine of 1769 epitomized this new age and led to a revolution in social life which we now know today as the Industrial Revolution. It was this revolution which underpinned Britain's ascendancy as an imperial, industrial power in the nineteenth century and established industrialization as a new goal of national focus for the European nation-states. While industrialization continues to be a key characteristic of scientific-technological cultures, these cultures are experiencing two further, technologically driven, revolutions. These are the Information Technology Revolution associated with advances in computer and electronic engineering and the Biotechnology Revolution associated with advances in genetic engineering and the possibilities emerging from the human genome project.

Four further features of scientific-technological cultures might be noted as indicated through the process of industrialization. Firstly, the Industrial Revolution expanded the production of surplus beyond that of food to all areas of material need. Fundamental to this revolution was the rise of the mechanical and extractive industries – the former providing the manufacturing and transport requirements for the new revolution, the latter providing the natural raw resources. Secondly, the creation of wealth from the new revolution required access to both expanding markets and a continuing supply of natural resources. As with the much earlier agrarian revolution, the new revolution gave rise to the emergence of new empires whose viability significantly depended on the acquisition, and exploitation, of new lands procured through a process of colonization. The process of colonization generated further requirements, especially militarily requirements, which provided a further stimulus to technological growth. Thirdly, central to the new revolution was a new system of financing the production and distribution of surplus, the system of capitalism. Fourthly, these previous three features, when taken together, generated a further dynamic to scientific-technological culture and that was that the culture should become an all-encompassing global culture. Scientific-technological culture now became a transformative goal for all societies (Noble, 2005). Accordingly, the variety, particularity and richness of traditional cultures were subsumed by the new culture with its universal and hegemonic interests.

In terms of the historical scenario provided we might also note that the technological inventions up to and including the Industrial Revolution had primarily sought to exploit the natural environment as the focus of agriculture and engineering and as the raw source of artifacts and armaments. However, the two further revolutions initiated in the latter part of the twentieth century, the Information Technology Revolution and the Biotechnology Revolution, have focused on two further environments for manipulation. One is that environment constituted by the world of human meanings (human information, knowledge, entertainment, media and culture), the other is that environment which constitute humans themselves as embodied, living organisms. While the human body has long been an object of interest, intervention and manipulation, biotechnology allows for the selection, shaping and manipulation of human bodies through control of the fundamental life (genetic) processes of the human individual. Both of these revolutions have been financed largely as entrepreneurial projects and the success of the projects necessitates that both human meanings and human bodies are largely redefined as commodities within a social setting of market exchange.

As noted previously, the change from hunter-gatherer to agrarian cultures led to changes in the relationships between settled communities and the land and between communities themselves. These changes in the environmental and socio-political conditions of human existence have grown in magnitude with the advent of scientific-technological culture and its three subsequent revolutions. The Industrial Revolution with its mechanization of agriculture and the craft industries saw a major shift in populations from the country to the new industrial cities, a transformation in wealth distribution and class structure, an emergence of urbanization and, at the international level, an aggressive competitiveness between the leading industrial states. The social changes introduced by that revolution have been significantly altered by the Information Technology Revolution which in the space of a few years has dramatically transformed the workplace, education, communication, entertainment and global landscapes (cf. Burton-Jones: 1999; Jones: 1995).

In their study of the history of technology from earliest times until the end of the nineteenth century Derry and Williams proposed that technological developments tended to affect human affairs in three significant ways:

The fortunes of mankind (sic) have been closely affected by the growth of technology; the same influence has helped to shape the relations of nations and classes; and in the life of the individual, technology plays an alternate role as servant or master of man's (sic) activities. (1960: 704)

They suggested that the most enduring features of technological cultures have been their growth in population numbers, their increasing mastery of more efficient and effective means of transport, their increasing spread across the surface of the globe and their use of struggle in the pursuit of power (1960: 704-708) They also noted that by the end of the nineteenth century three further features were emerging (1960: 708-711). Firstly, the complexity of technological enterprises had resulted in the growth of large units of production which required an administrative managerial bureaucracy who exercised a mediating role between the owners of capital and those who actually produce the goods. Secondly, these cultures have given rise to a workplace environment which in terms of psychological satisfaction has often proved to be alienating. Thirdly, they proposed that an emerging dynamic in shaping the further development of technology was not the satisfaction of long-felt wants, but the actual creation of new wants. In evaluating the worth of the technological enterprise, they made the following comment:

But to the more searching question, whether technological progress has on balance added to the happiness of the individual, we can at best offer no more than an affirmative answer hedged with qualifications, rejecting the temptation to pretend to weigh imponderables. (1960: 710)

A significant feature of many contemporary scientific-technological cultures is the change from an economic framework of industrial capitalism to one of consumer capitalism, however questions regarding the alienation of the workplace and the contribution of the new technologies to enhancing human happiness and wellbeing are still matters of concern to many (Eckersley, 2004; Hamilton and Denniss 2005; Sennett, 2006; Hamilton, 2006).

THE ETHICAL IMPULSE – MORAL VOICES

I have suggested that technology both shapes, and is shaped by, history. The same applies to the ethical impulse. Moral understandings and the ethical impulse have both shaped, and been shaped by, history and although that history is quite a complex one, some understanding of it is helpful in better making sense of the interface between technology and ethics. A convenient way of understanding the key features of the Western moral tradition is provided by the metaphor of the moral voice, that locus of moral authority to which humankind has turned at different times in history. Broadly speaking, that locus has changed with changing cultural circumstances (especially with those circumstances introduced by technological inventions). The long history of Western civilization gives rise to a succession of authoritative moral voices – the voice of nature, the voices of the gods, the voice of God, the voice of Reason and the voice of the Other.

From nature to the voices of the gods

Hunter-gatherer communities had a close, trusting relationship with the land which was seen as nourishing and protecting the tribe provided it, and the other living creatures within it, was treated with appropriate reverence and respect. Central to the life of the hunter-gatherer was the desire to live in harmony with the land and to live according to its voice. For the hunter-gatherer the primary moral voice was the voice of nature.

The move from a hunter-gatherer culture to an agrarian culture gave rise to new forms of human relationships, human social organization and a new relationship with the land. While the exploitation of the land by agricultural means was generally successful, humans became acutely aware of the vulnerability of such a settled lifestyle before the forces of nature. These forces were now seen as powerful, unpredictable and at times malevolent. Nature, since it could both nourish life and destroy it, was now represented as fractured, a site within which the interplay of the forces of good and evil was a never-ending struggle. These forces were now portrayed as the work of invisible deities, gods who create, nourish and also destroy and humankind was now seen as dependent upon the power and pleasure of these gods. Thus, early settled communities came to see their localities as governed by divine spirits who intervened in human affairs through the forces of nature and they sought to appease these forces in nature through prayer, ritual and sacrifice.

Scholars have suggested that the growth of settled communities into cities and city-states required a more complex level of social organization which further depended upon a commitment to a system of more complex moral values and dispositions of civility and cooperation. The extension of social membership to include others-as-strangers made possible even more complex social units which in turn made possible the generation of more, and more complex, social benefits in terms of material goods, artifacts, buildings, engineering projects and so on. The wellbeing of the city was now seen as dependent upon three key factors: harmony, civility and cooperation between its members; the ability of the city to protect itself from external aggressors; and the ability of the city to propitiate the forces within nature. A divine order was accepted as the anchoring point of stability and accordingly these early communities were theocratic, the gods worshipped being the personification of forces within nature. The temple became the focal point of village and town life, the priests were both rulers and administrators and religion became the fabric that bound the people together in a shared morality, in a shared identity and in a shared sense of meaning and purpose. Hence, the moral order, the social order and the natural order were all now seen as an extension of a divine order with the voice of the divine being interpreted by the ruler-priests. J. M. Roberts' observations regarding the ancient gods of Sumer could apply equally to any of the early human communities:

> These gods demanded propitiation and submission in elaborate ritual. In return for this and for living a good life they would grant prosperity and length of days, but no more. In the midst of the uncertainties of Mesopotamian life, some feeling that a possible access to protection existed was essential. Men (sic) depended on the gods for reassurance in a capricious universe. The gods – though no Mesopotamian could have put it in these terms – were the conceptualization of an elementary attempt to control environment, to resist the sudden disasters of flood and dust-storm, to assure the continuation of the cycle of the seasons by the repetition of the great spring festival when the gods were again married and the drama of Creation was re-enacted. After that, the world's existence was assured for another year. (J. M. Roberts, The Penguin History of the World, pp. 53-4)

The emergence of civilizations or empires first appears with the Sumerians in lower Mesopotamia from about 3500 BCE. Religion remains central to the lives and destinies of people within these empires. However, changes in social organization lead to changes in religious belief with the transition from a local god associated with the locality of a town or place to a comprehensive pantheon of gods ruling the universe and associated with humankind through a complex system of theology, ritual and priesthood. Religion still serves the same function of explaining dysfunction in nature and human affairs and of offering some further means, albeit religious means, whereby people might seek to control the unpredictable

The voice of God

The polytheistic religious tradition came to be challenged in the Near East by the monotheistic religions of Zoroastrianism, the ancient religion of Persia and named after the founding prophet Zoroaster or Zarathustra of about 1200BCE; Judaism founded in the near-east by the patriarch Abraham in approximately 1600BCE, but primarily dating from the time of Moses in about 1000BCE; and Christianity, the religion of the followers of Jesus of Nazareth, a Palestinian Jew who lived from 4BCE – 27 CE. It is probable that Zoroastrianism influenced Judaism which, itself, was the religion from which Christianity emerged

Each of these religions taught that there was only one God, not many gods, and that nature was, like humankind, part of God's creation. The orderliness, regularity and munificence of nature were now seen as the work of the Creator and God was now seen as revealed both through revelations to chosen individuals and through the works of nature. Hence, obedience to God's voice required obedience to both the laws of nature and those laws divinely revealed – the voice of God spoke through both the Sacred Books (the Scriptures) and the Book of Nature.

Furthermore, these three monotheistic religions each promoted a linear view of a salvific history. They believed that history began with the creation of the world as a paradisiacal place of peace and harmony. This creation is marred by the introduction of evil (depicted in the Judaeo-Christian scriptures by the 'fall' of Adam and Eve under the influence of an evil spirit, Satan) which leads to humankind being banished from paradise and having to live in a world affected by the consequences of sin – a world of work, illness, suffering, violence, catastrophe and death. However, in spite of this sin the world is not forsaken by God, but becomes the site of a new relationship between God and humankind whereby over time both humankind and nature can be restored to the original peace and harmony of paradise. Thus human history is linear in that, through the grace of God, it is moving towards a time of reconciliation, redemption and salvation. For the Jewish people this history has been inaugurated already in the covenantal encounters between Yahweh and his people (the encounters mediated with Abraham, Noah and Moses). By the time of the birth of Jesus of Nazareth there is a strong expectation among many Jews of a coming messiah who will realise the 'time of salvation'. For the followers of Jesus that time had come in the person of Jesus. For them Jesus is the one in whom the reconciliation between the divine and the fleshly has been achieved. Jesus, the god-man, is the first-born of the new, restored, creation and the harbinger of the restoration of the whole of creation.

The first followers of Jesus anticipated that the new creation, the second coming, would occur within their own lifetimes. When this did not occur, the conviction grew that such a time would come only when all people had come to believe in Jesus. This conviction saw the fledgling community break away from Judaism and endeavor to secure converts across the whole of the Roman Empire. While Christianity's success was rather gradual in the following centuries, under the auspices of the emperor Constantine in the fourth century CE, it became the de facto imperial religion of Rome. Across the empire the older local rituals and religions associated with the indigenous, pagan religions were gradually absorbed as over the following centuries Christianity became more powerful. Thus, in the Western cultural tradition the voice of God as interpreted within Christianity became the overriding moral authority from the time of the emperor Constantine in the fourth century until the cultural revolution of modernity which appeared on the European stage in the seventeenth century (Kung, 2003).

From this powerful religious tradition certain assumptions emerged regarding the human condition. Nature was seen as a created order, distinct from God and from humans, but provided for the benefits of humans. Nature was also seen as being governed by universal laws, the natural law, as consistent with God as the supreme logos. Since nature had been created for the benefit of humankind, it was there to be exploited. Although humankind was embedded in the natural world, more importantly, it was embedded in a sacred history that promised, ultimately, paradise. Accordingly, the moral form of life was to be understood as a struggle that unfolds over time and against the backdrop of this cosmic salvific history. History was sacred since it moved towards an endpoint in time, the time of salvation. At both the personal and cosmic levels a developmental process was in play, an inexorable historical process of redemption, reconciliation and final fulfillment. It was also a history whose meaning was ultimately other-worldly. The created natural/bodily order was seen as a transitory impediment to the ultimate fulfillment of the self as a spiritual being. Nature was to be seen as inferior to the spiritual and natural or base desires were seen as a threat to the life of the spirit. Accordingly, personal salvation involved a continuing struggle of the spirit over the desires of the flesh, and in this struggle both faith and reason could enlighten the moral path.

The Voice of Reason

In the seventeenth century a new cultural revolution – religious, intellectual, social, political and, finally, economic – took shape and over the succeeding three centuries swept across Western Europe. Disenchantment with religious authority, especially papal authority, and revulsion for the civil violence that marked the struggle between the old Catholic, or Roman, faith and the beliefs of the dissenters, or Protestants, led the intellectuals to seek a new foundation for moral and political authority. Technology had a crucial role in furthering the revolution, not least of all the new printing technology (Eisenstein, 1983). The revolution sought to sever ties with the past and to rebuild the edifice of human knowledge and the human social order de novo on a human, yet certain foundation. The new, certain authority was that of Reason, and it was the application of Reason, as first practiced in the rational inquiry of the Greek philosophers and as also emerging from the new 'natural philosophers' who sought knowledge through observation and experiment-ation, that was now postulated as the anchoring point of a new human civilization.

As Schneewind (1998) has shown, within the sphere of moral theorizing the displacement of the religious heritage by Reason is a gradual and complex process. The process reached its zenith in the works of Immanuel Kant (1724-1804). The Kantian moral program emphasized the place of individual decision-making as constitutive of the moral form of life. Right action reflected right judgment and right judgment reflected right Reason, i.e., Reason as exemplified in necessary and universal principles or laws. Kant's moral theory provided both an account of the good life in substantial terms (life as lived by a free or autonomous self) and in procedural terms (life as lived according to the requirements of the categorical imperative). Within the social order persons act rationally when they seek to confirm their wills with the requirements of the universal moral law, the categorical imperative. Within the order of nature persons act rationally when they seek to confirm their wills to the requirements of scientific laws. While a number of subsequent philosophers sought to articulate a different theoretical setting for the exercise of human autonomy, and with it different moral laws or principles, Kantian philosophy had defined the key characteristics of morality which today are still influential in shaping the conception of ethics and the good life within English-speaking philosophy, especially the discipline of applied ethics. These characteristics are that morality is concerned with the exercise of autonomous, individual agency, that such agency should conform to the requirements of Reason as displayed by universal moral principles, that all persons should be given dignity and respect inasmuch as they seek to act autonomously, and that the good life consists in shaping one's own life by maximizing the exercise of autonomy.

The rule of Reason in Western societies has flowered since the seventeenth century nourished by the growth of the scientific tradition, the related successes of technology and the emergence of a scientific-technological culture as indicative of a rational, progressive society as against the backwardness of those societies still wedded to superstition or religious beliefs. From an ethical perspective, two developments are worth noting. First, the foundational values of public life moved from a religious to a secular basis and, secondly, the meaning and purpose of human life moved from a focus on the transcendental (the life beyond) to a focus on the immanent (our here-and-now earthly life). The ultimate goal of modernity emerges clearly in the nineteenth century as the creation of a secular earthly paradise under the guidance of Reason – rational inquiry, rational understanding and rational processes. Within liberal, democratic and capitalistic societies this earthly paradise is to be secured primarily through understanding, dominating and exploiting the conditions of our immanence – the conditions of our being free and rational beings who are also embedded in the natural world. It involves, on the one hand, the pursuit of greater liberation from the constraints of the world (both material and biological), and, on the other hand, a greater opportunity to exercise wider choice in enjoying the benefits provided by that world. By the late twentieth century science and technology, the fruits of Reason, are seen by many as delivering a new world of promise, freedom, life and happiness.

TECHNOLOGY AND MORALITY – EXPLORING THE INTERFACE

The historical overview which I have provided seeks to illustrate in broad terms how technology has been pivotal in shaping the evolution of human cultures and the human condition. It also points to the way in which Western moral frameworks have also evolved, an evolution tied to the broader cultural changes which have been noted. What is not so apparent is the relationship between the cultures and the moral frameworks. Indeed, I would like to suggest that the moral frameworks have been a positive force in legitimating the technological imperative. In the main, the key moral orientations within Western culture have served to reinforce the validity of technology rather than subject it to critical challenge.

The move from hunter-gatherer communities to agrarian communities involved a change in the dynamics of human relationships – relationship to the land and relationship between communities. The dynamics of partnership and cooperation central to hunter-gatherer communities was replaced by a dynamic of domination and control (Brody: 2001) and the old moralities based on the Voice of Nature were seen as irrelevant. The natural losses and disasters which occurred from this new dynamic were rationalized away by the new religions and their mythologies as the acts of the gods and spirits. In effect, humans were not victims of their own technologies and social systems, but of malevolent forces within the world. Indeed, religion sought to counter these forces by extending the dynamic of control, or at least manipulation, to the realm of the gods. Prayer, sacrifice and divination evolved as 'spiritual technologies' whereby the favor of the gods and spirits might be guaranteed. Furthermore and as already noted, the good life was one of agricultural and military strength and of access to lifestyle goods. Hence, technology, and technological inventiveness, was central to the conception of human wellbeing that emerged with agrarian communities

The spread of Christianity across Western Europe after the time of Constantine supplanted the older pagan religions. However, Christianity adhered to the view that dysfunction within nature had its origin in the spiritual realm. It was the result of the introduction of sin into creation through Satan's seduction of the first people, Adam and Eve. The renewal of the world was primarily seen as a renewal of humankind morally and spiritually. For the first one thousand years of Christianity the spiritual pursuit of perfection disdained the mundane technology of the artisans. Under the influence of Platonism and asceticism Christianity emphasized spiritual renewal and preached the attainment of happiness in the other-worldly life after death. Within such a worldview little emphasis was given to the value of 'earthly concerns' and the world, both social and natural, was seen as governed by a hierarchical order that had been divinely ordained (Kung, 2003). Within such a religious ethic, the value of technology, as with all 'worldly possessions', is treated neutrally. It is taken as a given in a life that, while also given, is merely a preparation for the true life to come.

However, as Noble (1999, 2005) has argued, from the early Middle Ages another tradition emerged in Christianity which valued technology as a means of recovering that perfection lost in the fall of Adam. Technology came to be seen as an instrument of salvation and that salvific role gained greater emphasis with the passage of time and the growth of technological inventiveness. Technological progress was sanctified and legitimated as humankind's participation in God's restoration of creation through salvific history. The Christian theology of history provides both the justification for, and ultimate goal of, the technological imperative. Noble notes:

> (t)he religious roots of modern technological enchantment extend a thousand years further back in the formation of Western consciousness, to the time when the useful arts first became implicated in the Christian project of

redemption. The worldly means of survival were henceforth turned toward the other-worldly end of salvation.....The legacy of the religion of technology is still with us, all of us. Like the technologists themselves, we routinely expect far more from our artificial contrivances than mere convenience, comfort, or even survival. We demand deliverance. (1999: 6)

It is interesting to note that in much of public discourse contemporary technology is still portrayed in quasi-religious terms. Notions of struggle, fight and battle, the language of promise, hope and trust, confidence in the inevitability of the new age that will see the conquering of 'evil' and the further dissemination of 'good', and the depiction of the scientist/technologist as apart, as hero and as savior, all reflect features whose origins lie within the Christian theology of history.

The increasing role of reason in Western life, the advent of secularization, the gradual demise of Christianity as a force in public life and the emergence of a secular morality based on Reason has not proved to be an impediment to technological advancement. Indeed, the logic of technological growth has been seen as largely consistent with the logic of modern morality. As we have seen the morality that emerged from the modernity project valued the exercise of autonomous, individual agency, conformity to the requirements of Reason (i.e. rationality), and valued the maximisation of those conditions which allowed persons to shape their own lives. And technology has espoused similar values, especially from the nineteenth century. It has sought to anchor further development rationally on the understandings and critical processes provided by science. The alliance between technology, capital and consumerism has provided for the increased production of goods, the increased distribution of such goods and the increased access of people to the benefits of these goods. Modern technological developments have drawn from scientific rationality, have sought to further alleviate obvious situations of human suffering, and have sought to promote greater human freedom and choice. While, as with any human practice, human participants can subvert and distort the practice, in its authentic forms modern technological practice presents as the epitome of a modern, rational and moral human activity.

Two features of modern morality are especially pertinent and both can be found in Immanuel Kant's influential philosophy. Kant drew a distinction between our conditions of interiority (the noumenal self, that is, the inner self of reason, autonomy and freedom), and our conditions of exteriority (the phenomenal self, the self as body and as governed by the laws of science). Thus, for Kant the good life was one which was lived in accordance with the requirements of rational autonomy and the requirements of scientific laws (Coplestone, 1964; Schneewind, 1998). This distinction between the conditions of interiority and the conditions of exteriority has given rise to two influential trajectories in shaping the subsequent ethical sensibility of Western society and with it an influential conception of the good life. Firstly, ethical emphasis has turned towards the individual self and the good life has come to be seen as the realization of that self through choice, freedom and self-actualisation (Taylor, 1989; 1991). Secondly, the conditions of our exteriority, as governed by scientific necessity, have been expanded beyond natural phenomena to include history, technology and the market (Noble 2005).

While this expanded notion of scientific necessity is shaped by both the religious notion of salvation history and the nineteenth century positivist conviction that science reflects the summit of human knowledge, its consequence is that the good life, as rationally conceived, requires, not just a grudging toleration of modern technology and the modern market, but a positive commitment to their beneficent value and inexorable progress.

As we enter the twenty-first century and seek to grapple ethically with the conception of the good life that the long history of technology, and the shorter history of the market, has bequeathed us, we face a critical challenge in that our traditional conceptions of the ethical may well fail to provide an appraisive framework for evaluating the claims of contemporary technology. The Voice of Nature that speaks of a hunter-gatherer past and the Voice of God that direct us to the spiritual and the other-worldly are neutral in their power to transform the present. The Voice of God that speaks of an Adamic restoration of a new Eden, a new earthly paradise, or the Voice of Reason that promotes the progress of Western rationality and civilization and the ultimate attainment of an earthly utopia, is too intertwined with the scientific-technological-market culture to provide a detached and critical perspective. If there is to be a way forward, perhaps it lies in listening to the voice(s) of those who stand outside the established ethical traditions, especially those who recognise that the task of ethical critique requires, firstly, a reappraisal of what it is to be a human self, a human being and what conception of the good life is appropriate to such beings not as fallen sinners, not as purely rational beings, and not as beings whose conditions of exteriority is totally beyond their influence and control.

ACKNOWLEDGING THE POST-MODERN CONUNDRUM

Given the long historical moral legitimacy of the technological imperative, how is it as we enter the twenty-first century many now argue that the technological enterprise is morally problematical. What is it in contemporary technology that has changed that has brought about this widespread concern? While I would agree that a number of developments within contemporary technology give rise to ethical concern (ethics *in* technology), the fundamental crisis of confidence lies in the loss of a belief in the intrinsic ethical worth of the technological movement (the question is one regarding the ethics *of* technology). And here the key point to be made is that this loss is not primarily due to a radical change in the technological dynamic. Rather, it is due to changes emerging in ethical understanding and orientation. The ethical framework which is now emerging is neither a religious framework, a Christian framework, nor a modern framework, even though it may incorporate characteristics from each of these preceding approaches. This emerging framework displays a post-modern sensibility and within this framework the technological enterprise is seen as being problematical. In post-modern morality the nexus between technology and morality that has sustained, and legitimated, the technological imperative since the early days of agrarian settlement has been broken.

Post-modern morality - Acknowledging the Voice of the Other

Many scholars are now proposing that any conception of morality and the good life must take account of our immanence as human beings, or to use the more prevalent term within the moral literature, our embeddedness as human beings. There are many features to this embeddedness, but we might note the following (Priest, 1998; Abbey, 2000):

– we are each embedded in our own biographies, our own patterns of understanding, our own virtues and vices, our own interests, our own hopes and fears – in short, we are each our own distinctive identities;
– we are all embedded in traditions, practices and cultures;
– we are all embedded in relationships, institutions, social settings and communities;
– we are all embedded in nature, physically, biologically and ecologically.

This embeddedness provides both the source and the contours of our be-ing and be-coming.

The emphasis in modern morality on autonomy and freedom has emphasized the first realm above while science and technology have focussed primarily on the fourth of these realms. The significance of the second and third realms in shaping the good life has been largely overlooked. Indeed, given that contemporary public life is significantly shaped by modern morality and technological developments, a fundamental issue of concern for post-modern morality is that these two social forces serve to marginalize what is to be valued within these two intermediate realms. A morality based on epistemic rationality and individual choice has neither the metaphysical frameworks nor the appropriate value-systems to understand and assess human be-ing as historical, social and embodied. Such a morality offers a limited interpretive frame in assessing the meaning and worth of technological transformations especially in terms of the potential they offer for subverting or colonizing realms two and three above.

Realms two and three above situate human agency within relational settings. Human agency is not merely the possibility to do. More correctly, it is the possibility to do within, where the within draws attention to the social, relational settings of agency. Accordingly, moral decision-making needs to be understood, not as a process centered on individual, internal deliberations, but of collaborative negotiation which involves a fidelity to the guiding role of the general principles of morality held within our moral traditions and culture, an acknowledgement of the particularities of the participants, their relational commitments and their contexts of moral choice, and a fidelity to processes of negotiation which reflect integrity, inclusiveness and authenticity. Because moral engagement is socially embedded, and the social setting may have been determined by traditions and power dynamics which discriminate, marginalize and exclude, it is often the case that such settings have a 'taken-for-granted' legitimacy. Modern morality has ascribed authority and legitimacy to those who are rational and free. In effect, it has silenced the voices of those who are disempowered, vulnerable, dependent and marginalised. Hence, an emerging disposition in post-modern moral engagement is a healthy 'suspicion' of that voice, whether 'rational', 'expert', 'elite', that would seek to make decisions for others, especially the powerless, while yet excluding their voices.

The notion of a moral self who can simply transcend time, history, culture, materiality and circumstance and choose to act freely according to the requirements of a transcendental realm of Reason is an illusion (Toulmin, 1992). True, each person is a unique self, with a unique identity, who brings a unique 'one's own voice' to the moral endeavor. Yet each person is also inextricably connected to others and to the world, and this connection of self-with-other requires that moral engagement is first and foremost a meeting of voices, an authentic dialogue that is inclusive of the voices of all concerned participants. It also requires that moral engagement is not necessarily an engagement whose goal is that of a person's, or a society's, control over their surrounds, whether social or natural. Ethical engagement and technological interventions are both about the exercise of power. Since the emergence of agrarian communities, the dominant conception of power that has directed these two forces is that of power as domination and control. Post-modern ethics with its emphasis on relations, collaboration, nurture and care resurrects the much older conception of power as collaborative and cooperative, what Rollo May has referred to as nutrient and integrative power (1974). Such a relational ethics sees power, not as a dynamic which is to be exercised primarily over or against others, but as a dynamic to be exercised for, and with, others.

If this is how moral engagement presents itself to us in the twenty-first century, a morality of connectedness, of 'being-in-relationships', then it requires a willingness to enter humbly and openly into a dialogue with others. It requires that we not seek to impose our stories and values on others, but are willing to listen to, to accept and to engage with the voices of others. This is particularly a responsibility when we encounter those who are usually silenced in our society and in institutional settings, the weak, the vulnerable, the different, the marginalised and the suffering. Such a morality critically challenges many of the present practices within contemporary technology, particularly biotechnology. Many of the voices which command in this practice are either exclusive voices, such as 'the voice of the government' or 'the voice of the expert', or nebulous voices such as 'the voice of progress', or 'the voice of society'. Such voices are particularly discernible in that they claim, or are given, overriding moral authority, reject dissenting voices and the need for dialogue, and portray such dissenting voices as misguided, ignorant or even demonic.

CONCLUSION

Contemporary developments in biotechnology are an outcome of a long history of human changes, yearnings and aspirations. Central to this history within the Western cultural tradition has been the myth of Eden, the myth of paradise, a place of peace and harmony free from illness, labour and suffering. Modernity and the process of secularization have shifted the site of this paradise from the hereafter to the here and now; have shifted power in the achievement of this paradise from the redemptive grace of God to the rational inventiveness of humankind; have shifted the notion of the good life from one of spiritual redemption and salvation to one of material well-being and prosperity; and have shifted the source of moral authority from the 'voice of God' to the 'voice of Reason'. However, while modernity has directed itself to the immanent, it has been unwilling to accept the possibility of human fallibility in that project. The myth perpetrated is that we could have confidence in the movement since ethically it was underpinned by a transcendent, powerful and non-human authority – in this new instance, not the authority of a transcendent God as the guarantor of salvation history, but the authority of universal Reason as the guarantor of progress and enhanced human wellbeing.

Post-modern morality holds that a fundamentalist interpretation of 'the voice of Reason' is as misplaced as a fundamentalist interpretation of 'the voice of God'. Discerning moral wisdom is a far more complex process that must accept, and proceed from, the conditions of our embeddedness. What we now have to recognise, and to live with, are three key truths. Firstly, that we, we humans, are primarily responsible for our own human condition – both in the sense of how it has historically evolved and in the sense of future possibilities. Secondly, that we humans are inexorably linked to, and dependent upon, each other and our world of nature. Thirdly, that we humans are fallible, limited and often exploitative. We are particularly limited in that we can only interpret our human condition and human possibilities through cognitive frameworks that are themselves human constructs. We have to live with the reality that there is no divine or transcendent rational source that can anchor with certainty the frameworks we accept and the actions which we pursue in the light of those frameworks. We humans are also often exploitative. Indeed, the embracing of dominating and controlling power as the means to achieving well-being has not only fuelled the technological imperative but has often resulted in both nature and vulnerable others being grossly exploited as the raw resources from which our progress might be extracted.

The way forward in enhancing moral engagement lies, in no small way, in creating spaces in which many voices can be brought together in dialogue and collaboration. This is a critical ethical challenge for contemporary biotechnology and contemporary society. The challenge is complex and imposing one. How do we ethically inform those who are engaged in biotechnological practice and how do we inform those who are excluded? How do we create public spaces within which these voices can meet and dialogue? How do we shape biotechnological agendas so that they are responsive, responsible and inclusive? Furthermore, if there is no inexorable movement towards paradise and technological change is not wholly progress but a movement of gains and losses, how do we discriminate between that which we wish to preserve and that which we are quite happy to cast aside? How, too, do we balance between benefits and losses? The most crucial question we face is this. How can modern morality, modern political structures and modern technological processes continue to be validated in a post-modern world?

Jared Diamond, in his extensive discussion of human societies that have ecologically collapsed (Diamond 2005) argues for a five-point framework of possible contributing factors. These are: environmental damage, climate change, hostile neighbours, friendly trade partners and a society's response to its environmental problems. He suggests that 'the fifth set of factors always proves significant' (2005, p.11) and in the final section of the book reflects on why some societies make disastrous decisions (2005, Ch. 14). Diamond considers how our cognitive frames of interpretation and our values system, as well as other irrational failures, have contributed to the demise of once flourishing societies. It might be pertinent to conclude with the observations that conclude that chapter.

> We should admire not only those courageous leaders, but also those courageous peoples...who decided which of their core values were worth fighting for, and which no longer made sense. Those examples of courageous leaders and courageous peoples give me hope. They make me believe that this book on a seemingly pessimistic subject is really an optimistic book. By reflecting deeply on causes of past failures, we too, ...may be able to mend our ways and increase our chances for future success. (2005, p. 440)

Our trust in the beneficence of the technological imperative reflects long and influential interpretive frames regarding our interpretations of the human condition, human wellbeing and the good life. Rather than continue to trust such influential interpretations, it is surely timely that we reflect deeply on them and courageously seek to appraise them against what we now see from our post-modern perspective as our core, enduring and sustainable ethical values.

REFERENCES

Abbey, R. (2000). *Charles Taylor*. London: Acumen.
Barbour, I. (1993). *Ethics in an age of technology*. San Francisco: Harper.
Braudel, F. (1995). *A history of civilizations*. London: Penguin.
Brody, H. (2001). *The other side of Eden: Hunter-gatherers, farmers and the shaping of the world*. London: Faber and Faber.
Burton-Jones, A. (1999). *Knowledge capitalism*. Oxford: Oxford University Press.
Coplestone, F. C. (1964). *A history of philosophy: Modern philosophy: Kant* (Vol. 6, pt. II). New York: Image Books.
Derry, T. K., & Williams, T. I. (1960). *A short history of technology*. Oxford: Oxford University Press.
Diamond, J. (2005). *Collapse: How societies choose to fail or succeed*. New York: Viking.
Eckersley, R. (2004). *Well & Good*. Melbourne: The Text Publishing Company.
Eisenstein, E. (1983). *The printing revolution in early modern Europe*. Cambridge: Cambridge University Press.
Hamilton, C. (2006). *What's left? The death of social democracy*. Melbourne: Black Inc.
Hamilton, C., & Denniss, R. (2005). *Affluenza*. Crows Nest: Allen and Unwin.
Hindmarsh, R., & Lawrence, G. (Eds.). (2001). *Altered Genes II*. Melbourne: Scribe Publications.
Jones, B. (1995). *Sleepers wake! Technology and the future of work* (4th ed.). Melbourne: Oxford University Press.
Kung, H. (2003). *Christianity*. New York: Continuum.
Kuusi, P. (1985). *This world of man*. London: Pergamon.
May, R. (1974). *Power and innocence*. London: Souvenir Press.
Nisbet, R. (1994). *History of the idea of progress*. New Brunswick, NJ: Transaction Publishers.
Noble, D. F. (1999). *The religion of technology*. New York: Penguin.
Noble, D. F. (2005). *Beyond the promised land*. Toronto: Between the Lines Press.
Priest, S. (1998). *Merleau-Ponty*. London: Routledge.
Pippin, T. (1999). *Apocalyptic bodies*. London: Routledge.
Pollard, S. (1971). *The idea of progress*. Harmondsworth: Penguin.
Roberts, J. M. (1995). *The penguin history of the world* (3rd ed.). London: Penguin.
Schneenwind, J. B. (1998). *The invention of autonomy*. Cambridge: Cambridge University Press.
Sennett, R. (2006). *The culture of the new capitalism*. New Haven, CT: Yale University Press.

Smith, A. (1993). *Books to bytes: Knowledge and information in the postmodern era*. London: BFI Publishing.

Taylor, C. (1989). *Sources of the self: The making of the modern identity*. Cambridge, MA: Cambridge University Press.

Taylor, C. (1991). *The ethics of authenticity*. Cambridge, MA: Cambridge University Press.

Toulmin, S. (1992). *Cosmopolis: The hidden agenda of modernity*. Chicago: The University of Chicago Press.

Urmson, J. O., & Ree, J. (1991). *The concise encyclopaedia of western philosophy and philosophers*. London: Unwin Hyman.

Peter Isaacs
Program in Ethics and Human Rights
Queensland University of Technology

CLARE CHRISTENSEN

11. RISKY FUTURES

Reflections on Scientific Literacy

Australia is now part of a global community being described by many sociologists (e.g. Beck, 1992; Giddens, 1990; Castells, 1996) as simultaneously a 'knowledge' society and a 'risk' society.

The knowledge referred to is primarily scientific and technological knowledge and the risks are seen to be associated with the use of new technologies and certain consumer products. Australians are now faced with increasing personal and collective decision making on issues which involve significant science content, for example, the safety of genetically modified (GM) foods, the health risks of extended mobile phone use, the disposal of hazardous wastes and the safety of living near high voltage power lines. How are they to judge these issues? Since these decisions affect the quality of life of many Australians, it can be argued that scientific literacy has become more important than ever before.

But what is 'scientific literacy'? This paper briefly outlines traditional and evolving uses of this term in science education and introduces a new perspective on scientific literacy arising from research into 'the public understanding of science'. I argue that this perspective suggests a new conception of scientific literacy which offers greater relevance for decision making on socioscientific issues. Discussion here centres on a comparison of the new perspective with current conceptions of scientific literacy and science teaching practices in Australian schools. Some recent science education curriculum initiatives in Queensland and Victoria are then examined in the light of this discussion.

THE 'KNOWLEDGE/RISK' SOCIETY

Because of the proliferation of information and knowledge in recent times, contemporary society is often described as a 'knowledge' society. According to Muller:

> we are entering a new form of society in which the social organisation of knowledge and the social organisation of learning are dramatically changing. Whether we are examining the economy, the polity or the realm of society and culture, knowledge as a form of symbolic capital increasingly becomes *the* central form of productive capital. [Italics in original]. (Muller, 2000, p. 43)

*Naomi Sunderland et al. (eds.), Towards Humane Technologies: Biotechnology,
New Media and Ethics, 173–186.*

To a large extent this knowledge consists of, or is based on, scientific and technological knowledge. At the same time the application of this knowledge is seen to have increased the risks faced by individuals and the human population as a whole (such as risks from radiation, GM foods and global warming), hence the parallel use of the term *risk* society (Beck, 1992). Beck observes that in this knowledge/risk society, debates around environmental issues are framed in terms of the formulas of natural science and that the risks we face 'initially only exist in terms of the scientific (or anti-scientific) knowledge about them'(1992, p. 26) He goes further to say that 'because the risks are imperceptible in most cases, they require the 'sensory organs' of science - theories, experiments, measuring instruments - in order to become visible or interpretable as hazards at all'.(1992, p. 27)

Giddens sees successful existence in this knowledge/risk society as:

> depending simultaneously on trust in proliferating expert systems on the one hand and a deepening reflexivity at both an individual and an institutional level on the other, as citizens increasingly monitor, question, demand justification and accountability from, and otherwise try to cope with a world of increasing uncertainty and risk (Giddens, 1990, p. 88).

Understanding risks framed in scientific terms and engaging in critical questioning of scientific knowledge or scientific experts require some measure of scientific literacy. Over the last 50 years the goals of science education have been framed in terms of scientific literacy. Because scientific knowledge, its applications and their risks are now so pervasive in our society, scientific literacy has assumed a new importance in people's lives and science education has a crucial role to play in preparing future citizens to make personal and collective decisions on socio-scientific issues.

If scientific literacy is now so important, how well are we preparing young people in Australia for their future in a knowledge/risk society? Are they equipped, for example, to make an informed decision regarding the safety of regular mobile phone use? This is a typical socioscientific issue, an issue involving significant science content and competing knowledge claims in the scientific community. Whether to eat GM foods is a similar issue impacting on people's lives in a personal way and some adults may also face decisions about using reproductive technologies. In addition, there are issues requiring collective decision making, such as the siting of toxic waste dumps, overhead power lines or mobile phone towers.

Surveys in most Western countries have found only between 5% and 15% of the adult population to be scientifically literate (Shamos, 1995). Although the methodology of these studies has been seriously criticised, particularly in relation to assessing only limited content knowledge, the levels of scientific literacy are generally agreed to be low enough to challenge claims of effectiveness by science education in relation to the general population (Shamos, 1995). These levels are particular cause for concern if the construction of society as a knowledge/risk society based on science and technology is accepted.

It can be argued that, more recently, 15 year old Australian school students performed reasonably well on the 2000 OECD (Organisation for Economic Cooperation and Development) PISA (Programme for International Student Assessment) survey of scientific literacy (the mean scientific literacy score for Australia students was close to 530 where the average was 500 and the highest score 550), hence there is no cause for concern (OECD, 2000). However, the recent DETYA (Department of Environment, Training and Youth Affairs) report, *The Status and Quality of Teaching and Learning of Science in Australian Schools* gives an account of science education in Australia which presents much more worrying findings (Goodrum, et al., 2000).

The authors of the DETYA report construct an *ideal* picture of the quality of science teaching and learning and compare this with the *actual* picture of school science education. Both actual and ideal accounts were constructed from reviews of national and international literature, professional standards documents, National, State and Territory curriculum documents, case studies of best practice, teacher and student surveys, and focus groups of teachers, scientists and community members. According to this research, at secondary level students experience disappointment because:

> [t]he science they are taught is neither relevant nor engaging and does not connect with their interests and experiences.....disenchantment with science is reflected in the declining numbers of students who take science subjects in the post-compulsory years of schooling (Goodrum, et al., 2000, p. viii).

The DETYA Report authors argue that the report is a serious indictment of any claim that current science education is producing scientifically literate citizens. This report placed a strong priority on students' engagement with science in the real world and it concludes that this goal is not currently being met. Thus scientific literacy is an issue of concern in the education of young people in Australia today.

CONCEPTIONS OF SCIENTIFIC LITERACY IN SCIENCE EDUCATION

How has the term 'scientific literacy' been used in science education? Scientific literacy was first used in the 1950s (DeBoer, 1991) and became the goal of science teaching in the 1960s, but it was taken to mean primarily content knowledge and, to a lesser extent, the processes of science. Science courses did not include applications of science or links to the daily lives of students. This was primarily because the aim of science education was seen as the training of future scientists, engineers and others such as those working in the health professions.

From the 1960s through to the 1990s the use of the term scientific literacy prevailed as the goal of science education, but in the 1980s critique of the term also emerged and alternative definitions of scientific literacy began to appear. For example, Shen (1975) and Branscomb (1981) proposed distinct kinds of scientific literacy, Shen naming three: practical, civic and cultural, and Branscomb eight categories which were essentially 'functional' literacies related to specific contexts (e.g. for scientists at work, citizens in daily life, journalists writing about science). Shamos (1995) suggested that 'scientific awareness' is a more useful term. By this term Shamos meant 'awareness of the enterprise, that is as a *cultural* imperative, and not primarily for content'(p. 217). He called this scientific awareness a 'functional literacy', and protested that the science education community keeps insisting on traditional scientific literacy, or knowing textbook science, instead of aligning science education with true public needs.

More recent conceptions of scientific literacy in Australia (Goodrum et al., 2000, p.165) and the UK (Millar and Osborne, 1998), place emphasis on coping with science and scientific information in a meaningful way, and having the interest and confidence to do this. In the Australian DETYA report, Goodrum, Hackling and Rennie noted that discussion about scientific literacy now generally focuses on three areas: the content and concepts of science, the nature and processes of science and the relationship between science and society. They consider a citizen to be scientifically literate if he/she is:

– interested in understanding the world;
– able to engage in discourses of and about science;
– sceptical and questioning of claims made by others about scientific matters;
– able to identify questions and draw evidence-based conclusions; and
– able to make informed decisions about the environment and his/her own health and well-being. (Goodrum et al., 2000, p. 2).

In the UK *Beyond 2000* report Millar and Osborne argued that 'School science should aim to produce a populace who are comfortable, competent and confident with scientific and technical matters and artefacts. The science curriculum should provide sufficient scientific knowledge and understanding to enable students to read simple newspaper articles about science, and to follow TV programs on new advances in science with interest. Such an education should enable them to express an opinion on important social and ethical issues with which they will be increasingly confronted'. 'Comfortable' and 'confident' are new terms appearing in the definition of scientific literacy, reflecting a shift towards personal needs. The inclusion of ability to 'identify questions and draw evidence-based conclusions' is also a significant change because it places new emphasis on being able to apply understandings about the processes of science beyond the classroom, 'where an ability to sift, sort and analyse information is paramount'.(Millar and Osborne, 1998, p. 9)

The US long term *Project 2061,* sponsored by the American Association for the Advancement of Science (1989), aimed at reforming science education, espouses a broad conception of scientific literacy which enables individuals to 'deal sensibly with problems that often involve evidence, quantitative considerations, logical arguments, and uncertainty' (p. 13). Thus in the UK, US and Australia current conceptions of scientific literacy are similar in their focus on the role of scientific knowledge and knowledge about science in people's lives, particularly in their decision-making about personal and social issues.

Whilst the above discussion refers to different national framings of scientific literacy, perhaps the most significant recent definition of scientific literacy has been made by science educators involved in the OECD PISA Project. Contributors to the development of this project included leading science educators from 32 OECD (industrialised) countries, hence this conception can be assumed to represent a significant international consensus. The PISA Project involves a three-yearly (from 2000) survey of the knowledge and skills of 15-year-olds in the principal industrialised countries. It assesses 'how far students near the end of compulsory education have acquired some of the knowledge and skills that are essential for full participation in society. It presents evidence on student performance in reading, mathematical and scientific literacy, reveals factors that influence the development of these skills at home and at school and examines what the implications are for policy development' (OECD, 2000, p. 2). According to the PISA authors:

> Scientific literacy is the capacity to use scientific knowledge, to identify questions and to draw evidence-based conclusions in order to understand and help make decisions about the natural world and the changes made to it through human activity. (OECD, 2000, p.1).

The above discussion of the meaning and use of the term scientific literacy in the science education research literature has demonstrated a shift over the past 50 years from a focus on content knowledge towards placing more importance on, and making more specific, the aspects of science by which it is involved with society and with individual lives. However recent research in the area of the public understanding of science suggests that this shift is only a beginning towards defining scientific literacy in ways appropriate for future citizens in a 'knowledge/ risk' society.

TOWARDS A NEW CONCEPTION OF SCIENTIFIC LITERACY?

The consistent findings of recent research into the public understanding of science pose significant challenges to school science educators. In this literature, dimensions of scientific knowledge not usually considered in school science are foregrounded and appear to demand attention in the education of future citizens for a society of both increasing knowledge and increasing risk. These dimensions include the uncertainty of much scientific knowledge, the evaluation of evidence, judging the trustworthiness of experts and an entirely pragmatic conception of content knowledge. In this section landmark research revealing these dimensions of scientific literacy will be described and its findings discussed.

The most widely cited research in the public understanding of science literature is the Leeds' case studies by Layton, Jenkins, Macgill, and Davey (1993) conducted in the UK. The aim of these studies was 'to explore what adults perceive to be their needs for scientific knowledge in relation to concerns which they have defined and prioritised themselves' (p. 26). Thus the researchers' starting point was not what scientists thought was important scientific knowledge but what members of the public saw as problems in their lives where science had the potential to contribute to a solution.

Four separate case studies were conducted with participants who were: parents of Downs' Syndrome children, elderly people living alone coping with domestic heating problems, County councillors serving on a Waste Management Sub-Committee examining methane gas release problems at a local landfill site, and residents living near the Sellafield nuclear re-processing plant. The first three studies involved clearly-defined groups facing a common problem and the focus of the investigations was on their perceptions of available scientific knowledge. The fourth study involved representatives of diverse groups reflecting the social composition of the towns and villages near Sellafield, for example, workers at the reprocessing plant, housewives, retired people and public sector professionals. In this study interest lay more in the dynamics of the flow of scientific information between groups concerned with the issue rather than a particular group's perceptions of the scientific knowledge available to them. The findings of these four case studies by Layton et al. have since been strongly supported by much further research in this field, including a group of seven studies reported by Irwin and Wynne (1996).

The principal findings of the Layton studies will be briefly outlined here then discussed later in relation to current conceptions of scientific literacy and contemporary Australian science education.

The first finding of the Layton et al. (1993) studies was that: '[t]he represent-ation of science as a coherent, objective and unproblematic entity characterised by certainty and direct applicability to everyday life received little support from the case studies' (p. 118). Much of the scientific knowledge available to participants was not able to provide unambiguous answers to their expressed needs, undermining the notion of science as certain knowledge which is useful for solving real world problems. In cases where experts presented conflicting evidence or interpreted research findings differently, the incomplete nature of the scientific knowledge was revealed.

Secondly, the traditional cognitive deficit model of public understanding of science was challenged by these case studies. In this model of the public understanding of science the lay public are considered to be ignorant, lacking both knowledge of scientific concepts and the motivation to use science in everyday life. This deficit is considered to be remediable by education about basic science concepts, in a one-way flow from experts to recipients. Contrasting with this view, the case studies demonstrated that 'the lay recipients of scientific knowledge were far from passive; they interacted with the science, testing it against personal experience, contextualising it by overlaying it with particular local knowledge and evaluating its social and institutional origins'. Further, 'when science is seen as articulating in useful ways with their concerns and interests, people demonstrate considerable resourcefulness in locating sources and impressive capability in translating science and other knowledge into forms which support practical action'(Layton et al., 1993, p. 122).

The interaction of these adults with science was not simply a cognitive one; in some cases considerations of personal identity were involved – their attitude towards science was influenced by how they 'see' themselves and their place in the world. Perhaps as a result of the influence of this sense of identity, a corollary to the above finding was that much public ignorance, as implicit in the cognitive deficit model of the public understanding of science, may be a selective, functional choice, allowing attention to be directed elsewhere. In place of the cognitive deficit model Layton et al. propose an 'interactive' model of public understanding of science.

A third major finding of this group of studies was that science does not operate in the real world as the objective, authoritative body of knowledge, contrary to its usual representation as such in public discourse. In the researchers' words, 'people do not encounter scientific knowledge as free-floating and unencumbered by social and institutional connections. The questions, 'From whom?' and 'From where? From what institutional source?' are central to judgments about the trustworthiness and reliability of the knowledge' (Layton et al., 1993, p. 122). In many instances in the case studies, participants' responses to the science presented to them by experts were crucially affected by how they perceived the presenters, as sharing or not sharing their particular perspective on the problem at hand.

Finally, participants in the studies often used science but reworked it in a particular situation to create new knowledge. This was not a straightforward application of scientific knowledge, any more than technology is applied science. This observation challenges the commonly noted dichotomy between everyday knowledge and scientific knowledge and supports ideas of "situated cognition" (Lave, 1998). In this view, context is an integral part of knowledge and thinking, not a separate, external factor. The researchers in the Leeds studies drew parallels with engineering knowledge, which is a distinct form of knowledge which translates abstract scientific knowledge into practical actions.

As suggested in the title of their report, *Inarticulate Science? Perspectives on the Public Understanding of Science and Some Implications for Science Education,* in this widely cited group of studies, Layton et al. illustrated the barrier which separates the general public from scientific knowledge. They concluded that science is to some extent 'inarticulate' in relation to the broader community. At the same time, they reported positively the many ways in which adults in the community manage to interact with science and on the basis of this they propose their 'interactive' model for the public understanding of science. This interactive model not only offers a more adequate tool than the cognitive deficit model for analysis of the interactions of adults in the community with science, but also suggests ways to proceed in efforts to improve scientific literacy both for members of the public and students in schools.

According to the research findings described above, using science in the real world to make decisions on socioscientific issues involves three main skills: confronting the uncertainty of much recent scientific knowledge, evaluating scientific evidence and judging the trustworthiness of experts. These skills will be considered in the next section, in the light of contemporary conceptions of scientific literacy and science education practices in Australian schools. In the final section of the paper, attention to these dimensions in recent science education reform initiatives will be examined.

NEW DIMENSIONS OF SCIENTIFIC LITERACY

Dealing with uncertainty

A feature of adult scientific literacy described by researchers in the public understanding of science and other authors (for example, Jenkins, 1997; Thomas,

1997) is that when facing socioscientific issues adults confront a kind of science which is uncertain and often controversial. For example, scientists may disagree about the interpretation of experimental results, or commercially funded scientists may present findings in a different light from government funded researchers. Thus this kind of scientific knowledge is uncertain 'science-in-the-making' (Latour, 1987) or 'frontier science' (Bauer, 1992), in contrast to the apparent certainty of 'ready-made' or 'textbook' school science which is usually presented as well established knowledge, no longer debated in the scientific community. Contested scientific knowledge is not mentioned in current conceptions of scientific literacy and is rarely explored in science classrooms. Thus being able to handle competing knowledge claims is a new dimension of scientific literacy.

The ability to deal with uncertain scientific knowledge involves understanding that the production of scientific knowledge is a social process which includes different interpretations of evidence or different methods being debated within the scientific community. This kind of understanding is essential to understanding the controversies surrounding much contemporary science. Such preparation could be addressed by the inclusion of some study of the social nature of science in school curricula. Current practice in Australian schools relies largely on the traditional normative conception of a single 'scientific method', involving simply the sequential, rational steps of hypothesis, observation and data collection and interpretation of evidence leading to hypothesis confirmation or rejection (Goodrum et al., 2000). This view is now highly outdated, in the light of writings in the philosophy and sociology of science since the 1970s. The social construction of scientific knowledge has now been well described by studies in the sociology of science (for example, Latour, 1987; Gilbert & Mulkay, 1984) but this knowledge has not usually been included in science teacher training and is not readily available to practising teachers. However, this is a view of science which may challenge the belief of many science teachers in the objective authority of science and thus may not be easily accepted.

Evaluating evidence

Using science in the real world may involve the evaluation of scientific knowledge/evidence, although research in the public understanding of science suggests that this is not always the path chosen by laypersons. Although the skill of drawing evidence-based conclusions has been specified in recent science education conceptions of scientific literacy, the comparative evaluation or critical questioning of evidence is not traditionally included in the conceptualisation of scientific literacy. Tytler, Duggan, and Gott (2001) argue that to develop these skills requires experience of a wide variety of investigative work which, according to the recent DETYA report, is not a common approach to science in contemporary Australian classrooms. Practical work usually consists of simple recipe type activities which are likely to lead students to an oversimplified view of science and its certainty compared with 'authentic science inquiry' (Chin & Malhotra, 2002). In authentic inquiry students would consider experimental design and examine evidence, reconciling theory and data even when data are uncertain or conflicting. A variety of such authentic tasks is possible, ranging from hands on investigations in the local community to the use of databases, computer simulated experiments and evidence evaluation tasks using the Internet. This kind of task could reasonably be expected to lead to a sounder understanding of scientific epistemology and better preparation for the contested science of contemporary issues.

Judging the trustworthiness of experts

As reported in the public understanding of science research discussed above, many people choose to rely on expert advice in areas of scientific knowledge which they feel unable to understand. The use of experts usually involves a choice between experts, and thus an assessment of their relative trustworthiness as a source of scientific knowledge. Evaluating trustworthiness is certainly a new consideration for science educators and many would argue that it lies outside of their domain of responsibility. However, if scientific literacy levels do not improve radically and scientific and technological advances continue at the present rate, people will continue to rely on experts. It can be argued that judging their credibility requires ways of assessing it based on understanding something of the nature of scientific evidence, which *is* the province of science educators.

A useful perspective could include critical literacy, as described by writers in the field of language education. For example, Lankshear (1997) describes three potential *objects* of critique with respect to critical literacy. Two of these objects of critique are relevant here – particular texts and wider social practices which are mediated by, made possible and partially sustained by texts. In critical literacy practice, questions are asked of texts, such as: 'What version of events/reality is foregrounded here? Whose version is this? From whose perspective is it constructed? What other possible versions are excluded? Whose/what interests are served by this representation? How does this text position the reader? What assumptions about readers are reflected in the text?' (Lankshire, 1997, p. 54). These are the kinds of questions needing to be asked in any evaluation of the arguments (texts) of experts; thus arguably 'scientific literacy' could be re-framed as 'critical scientific literacy'.

Together these three aspects of using science - dealing with uncertain science, evaluating evidence and judging the trustworthiness of experts - constitute a radical shift in thinking from conceptions of scientific literacy used in science education from the 1950s to the 1980s. The four recent conceptions of scientific literacy from the US, UK, Australia and the international OECD PISA Project (described above) place much greater emphasis on using science in real world contexts than previous definitions. However I argue here that the public understanding of science research described above leads us beyond these conceptions, offering new dimensions of scientific literacy for the future.

In this new conception some understanding of the social construction of scientific knowledge is vital in order to deal with the uncertainty of 'science-in-the-making'. Being able to evaluate and question evidence are essential skills, and critical language skills are needed to assist in judging the trustworthiness of sources of scientific knowledge. Further, the role of content knowledge is no longer central, as adults in the UK studies (described above) amply demonstrated, proving themselves adept at finding content on a need-to-know basis. In short, to prepare young people for a knowledge/risk society in which they will face increasing socioscientific issues, this new kind of literacy needs to be addressed in science education in Australian schools.

Recent initiatives

Concerns about the effectiveness of current Australian science education expressed in the recent DETYA report (above) are being addressed in a variety of ways in most Australian States and Territories. Whilst it is beyond the scope of this paper to survey all of these reforms, some recent initiatives in Queensland and Victoria will be considered in relation to the dimensions of scientific literacy discussed above.

Nationally, there has been a sense of urgency in recent times regarding declining enrolments in Year 11 and 12 Physics and Chemistry and in tertiary science courses, particularly in relation to the future provision of adequate numbers of scientists and engineers in the workforce and Australia's competitiveness in the globalised knowledge economy, based as it is so strongly on scientific and techno-logical knowledge (Batterham, 2000; Committee for the Review of Teaching and Teacher Education, Department of Education, Science and Training, (2003). This has been a primary motivation behind recent initiatives in Queensland and Victoria to improve student engagement in science education. As argued above, however, of equal importance is the need to equip *all* future citizens with science understandings to prepare them for individual and collective decision making on socioscientific issues.

In 2003 the Queensland government commissioned a committee of experts from science, industry, education and government to produce a vision for Queensland science education for the next three years. The resulting *Spotlight on Science* document aims to provide the basis for a renewal of science education in Queensland (Education Queensland, 2003a). Although this initiative has arisen from political and economic imperatives, particularly the goal of positioning Queensland at the forefront of the 'knowledge economy' to become a regional leader in 'smart' industries, such as biotechnology, (Department of Innovation, and Information Economy, 2003, p.10) the document places some emphasis on outcomes of 'Science for Life' and 'Connecting Science'. Connecting science refers to teaching science 'as a continuing process of practical inquiry into real-world issues and problems'. 'Science for life' aims to enable young people 'to make responsible decisions concerning their own and the community's future wellbeing' (Education Queensland, 2003a, p.6). This decisionmaking is closely linked with the consideration of contemporary socioscientific issues and the vision statement thus focuses attention onto citizen's needs. If the 'real-world issues' referred to include such contested knowledge as the science of mobile phone safety, GM foods, high voltage power lines and other environmental problems, the direction of this vision supports concern about the aspects of scientific literacy addressed in this paper. As discussed earlier, however, an important component of dealing with the contested knowledge of contemporary issues is some understanding of the nature of scientific knowledge as socially constructed and consensual. In the *Spotlight on* Science vision, discussion relating to 'Connecting Science' notes that 'it is vital that science education also reflects the provisional nature of scientific knowledge' (Education Queensland, 2003a, p.10), but the importance and role of understanding the nature of science is not mentioned elsewhere in the document.

The vision is a broad statement rather than a curriculum document. It supports the concerns expressed in this paper to some extent, but to fully address the dimensions of scientific literacy discussed here would require a significant shift away from current science teaching practices, for example in the use of authentic inquiry as opposed to recipe type practical tasks and much greater emphasis on understanding the nature of science as consensual knowledge, and being aware of its strengths and limitations. In addition new teacher skills are needed to facilitate the wider ranging discussions that addressing contemporary issues would encompass.

An encouraging curriculum initiative in Queensland is the current trialling of *Context-Based Science* in Year 11 and 12 Physics and Chemistry (Beasley & Butler, 2002). Here context is defined as 'a group of related situations, phenomena, technological applications and social issues' and teachers are free to develop contexts based on local interests and contemporary socioscientific issues. In this approach the starting point for learning is 'authentic real world experiences' and greater emphasis is placed on learning content 'in the context of inquiry, technology, personal and social perspectives and the history and nature of science' (Beasley & Butler, 2002, p. 2). This contextual basis offers teachers an opportunity to incorporate contemporary socioscientific issues into Chemistry and Physics, alongside more predictable contexts like sport, cars and air pollution. It also has the potential to offer experience in the evaluation of evidence, but only if teachers engage students in authentic forms of inquiry rather than the traditional recipe approaches. Further, learning content in the context of current social issues is likely to involve examination of the links and interactions of science with society, in turn leading to greater understanding of the role of social and cultural factors in the construction of scientific knowledge.

The actions proposed in the *Spotlight on Science* vision aim to improve the professional development of science teachers and make more links with the community, particularly industries and scientists. This is similar to the major foci of another recent Queensland initiative, the *New Basics* curriculum (Education Queensland, 2003b) across Years 1 to 9, which has recently completed its first three years of trialling in 59 schools. In order to better prepare young people for the future, the 'old basics' (commonly known as the 3R's) have been replaced by a broader range of qualitatively different literacies. Whilst pedagogy (in the form of 20 'productive pedagogies') and greater interactiveness of schools with the community are major foci of the *New Basics* curriculum, equally important is a strong transdisciplinary approach to curriculum. Student work in the *New Basics* curriculum is organised around 20 transdisciplinary 'rich tasks'. In relation to the conception of scientific literacy discussed above this transdisciplinary emphasis is promising in opening up traditional science content to broader understandings of the nature of science and its interactions with society. Since the rich tasks are transdisciplinary, none of them focuses explicitly on science, except the Years 7-9 task, *Science and Ethics Confer.* This task offers, over a 3 year period, an opportunity for the discussion of contemporary socioscientific issues and the nature of scientific knowledge in middle school science classrooms. The innovative framework of the *New Basics* curriculum involves a major shift in the organisation of school science content and student learning towards more open inquiry and authentic tasks. In this way it offers excellent opportunities for the development of skills such as dealing with uncertain science, critically evaluating evidence and judging the trustworthiness of sources of knowledge.

As in Queensland, a Victorian school science reform programme called *School Innovation in Science* was initiated by the State government (Victorian Department of Education and Training, 2003). It arose from concerns about the future development of the state as a knowledge economy, but in addition stated explicitly a strong commitment to scientific and technological literacy for all citizens. As in Queensland a major focus of the reform has been on teacher professional development and the forging of links with a wide range of community groups including scientists, engineers and industries. This reform agenda appears to address some of the dimensions of scientific literacy discussed above. For example, in the central framework of the initiative, *Components of Effective Teaching and Learning in Science*, effective science teaching encourages students 'to question evidence in investigations and in public science issues'. Science is presented as having 'varied investigative traditions and constantly evolving understandings', with 'important social, personal and technological dimensions' and 'successes and limitations [which] are acknowledged and discussed' (Tytler & Conley, 2002). Further, links are made between the classroom and the community, which emphasise the social and cultural implications of science. If implemented, these dimensions of science learning provide students with opportunities to develop the skills of dealing with the uncertain science of contemporary issues and the critical thinking required to evaluate evidence and judge the credibility of sources of scientific knowledge.

These Queensland and Victorian initiatives provide grounds for optimism that science education may be able to meet changing personal and societal needs. This depends, of course, on history not repeating itself, as earlier 'Science Technology Society' reform efforts in the 1980s and 1990s failed to break the stranglehold of academic scientists over school curricula. This has been recently documented by leading Australian science educator Peter Fensham (2002), who has long advocated a *science for all* approach in science education, encompassing the needs of the majority of students as well as the minority who will proceed to science-related careers (Fensham, 1985).

CONCLUSION

This chapter has examined conceptions of scientific literacy in the light of contemporary Australian needs, in particular the need for citizens in an increasingly scientific/technological society to be able to make informed choices and decisions involving scientific knowledge. It has been argued that recent research into the public understanding of science foregrounds aspects of knowing science which are not a strong focus in current Australian science teaching and which are important if young people are to be prepared by their schooling to face increasing socioscientific issues, affecting, for example, what they eat, how they communicate and possibly how they reproduce.

It is argued here that contemporary socioscientific issues need to be made an integral part of school curricula. Through the discussion of such issues, including their ethical dimensions, school science can develop important skills which are needed in this context and which are currently neglected. These skills include: dealing with uncertainty in scientific knowledge, critically evaluating evidence and judging the credibility of sources of scientific knowledge. Recent science education initiatives in Queensland and Victoria have been briefly considered and constitute grounds for cautious optimism.

Little research has been done into how adults in the Australian community interact with science in their everyday lives, particularly in dealing with the contested knowledge of contemporary debates. Research is needed which would highlight the issues discussed above and focus attention upon the science education needs of laypersons, particularly in relation to personal and collective decision making. This could contribute in turn towards reforms in science education which could ultimately improve people's chances of understanding and dealing with the scientific knowledge that is so rapidly changing their world.

REFERENCES

American Association of the Advancement of Science. (1989). *Science for all Americans: Benchmarks for scientific literacy*. Retrieved from http://www.aaas.org/project2061

Bauer, H. (1992). *Scientific literacy and the myth of the scientific method*. Urbana, IL and Chicago: University of Illinois Press.

Beasley, W., & J. Butler, (2002, August 12–14). *Implementation of context-based science within the freedoms offered by Queensland schooling*. Paper presented at the Australian Science Education Research Association Conference, Townsville.

Batterham, R. (2000). *The chance to change: A public discussion paper*. Retrieved August 17, from http://www.isr.gov.au/science/review

Beck, U. (1992). *Risk society: Towards a new modernity*. London: Sage.

Branscomb, A. W. (1981). Knowing how to know. *Science, Technology and Human Values, 6*(36), 5–9.

Castells, M. (1996). *The rise of the network society*. Oxford: Blackwell.

Chinn, C., & Malhotra, B. (2002). Epistemologically authentic inquiry in schools: A theoretical framework for evaluating inquiry tasks. *Science Education, 86*(2), 175–218.

Committee for the Review of Teaching and Teacher Education, Department of Education, Science and Training. (2003). *Australia's teachers: Australia's future*. Retrieved from www.detya.gov.com.au/schools/teachingreview

DeBoer, G. E. (1991). *A history of ideas in science education: Implications for practice*. New York: Teachers College Press.

Department of Innovation and Information Economy. (2003). *Investing in science: Research, education and innovation*. Retrieved September 10, 2003, from http://www.iie.qld.gov.au/publications/smartstate/default.asp

Education Queensland. (2003a). *Science state smart state: Spotlight on science 2003–2006*. Retrieved September, 2003, from http://education.qld.gov.au/publication/science/sciencestate.html

Education Queensland. (2003b). *The new basics project*. Retrieved September, 2003, from http://education.qld.gov.au/corporate/newbasics/

Fensham, P. (2002). Time to change drivers for scientific literacy. *Canadian Journal of Science, Mathematics and Technology Education, 2*(1), 9–24.

Fensham, P. (1985). Science for all: A reflective essay. *Journal of Curriculum Studies, 17*, 415–435.

Giddens, A. (1990). *The consequences of modernity*. Cambridge: Polity Press.

Gilbert, N., & Mulkay, M. (1984). *Opening Pandora's box: A sociological analysis of scientists' discourse*. Cambridge: Cambridge University Press.

Goodrum, Hackling, M., & Rennie, L. (2000) *DETYA report: The status and quality of teaching and learning of science in Australian schools: A research report prepared for the Department of Education, Training and Youth Affairs*.

Irwin, I & Wynne, B. (VVVV). *Misunderstanding science? The public reconstruction of science and technology*. Cambridge: Cambridge University Press.

Jenkins, E. (1997). Towards a functional public understanding of science. In R. Levinson & J. Thomas (Eds.), *Science today: Problem or crisis?* (pp. 137–150). London: Routledge.

Lankshear, C. (1997). *Changing literacies*. Buckingham: Open University Press.

Latour, B. (1987). *Science in action: How to follow scientists and engineers through society*. Cambridge, MA: Harvard University Press.

Lave, J. (1988). *Cognition in practice: Mind, mathematics and culture in everyday life*. Cambridge; New York: Cambridge University Press.

Layton, D., Jenkins, E., Macgill, S., & Davey, A. (1993). *Inarticulate science? Perspectives on the public understanding of science and some implications for science education*. Driffied, UK: Studies in Education Ltd.

Millar, R. & Osborne, J. (1998). *Beyond 2000: Science education for the future*. London: King's College London, School of Education, Report of a Seminar Series.

Muller, J. (2000). *Reclaiming knowledge: Social theory, curriculum and education policy*. London: Routledge Falmer.

Organisation for Economic Cooperation and Development (OECD). (2000). *Programme for International Student Assessment Project (PISA)*. Retrieved from http://www.pisa.oecd.org/index.htm

Shamos, M. H. (1995). *The myth of scientific literacy*. Brunswick, NJ: Rutgers University Press.

Shen, B. S. P. (1975). Scientific literacy and the public understanding of science. In S. B. Day (Ed.), *Communication of scientific information*. Basel: Karger.

Thomas, J. (1997) Informed ambivalence: Changing attitudes to the public understanding of science. In R. Levinson & J. Thomas (Eds.), *Science today: Problem or crisis?* London: Routledge.

Tytler, R., & Conley, H. (2002, August 12–14). *The science in schools research project: Results and insights from a system wide change initiative*. Paper presented at the Australian Science Education Research Association Conference, Townsville.

Tytler, R., Duggan, S. & Gott, R. (2001). Dimensions of evidence, the public understanding of science and science education. *International Journal of Science Education, 23*(8), 815–832.

Victorian Department of Education and Training. (2003). *Science in schools strategy*. Retrieved September, 2003, from http://www.scienceinschools.org/home.htm

Clare Christensen
School of Cultural & Language Studies
Queensland University of Technology

NAOMI SUNDERLAND

12. BIOTECHNOLOGY, SOCIETY, AND ECOLOGY

Understanding Biotechnology as Mediation

INTRODUCTION

The purpose of this chapter is to provide a framework for understanding the ways that biotechnological interventions and the ways of seeing associated with contemporary biotechnology can influence both ecology and society at a macro level. In particular, this chapter seeks to highlight exactly *how* a practice such as biotechnology produces, reproduces, imports and exports distinct social representations and manifestations of how we should see, be, and act both as persons and as a species. Framing biotechnology as a series of processes of mediation allows us to analyse biotechnology not as a set of static, isolated techniques or technologies but, rather, as an inherently political means of producing, reproducing, and shifting meanings across contexts of both organic and inorganic production and consumption. I posit that the social practice of biotechnology extends outward into society, and in turn is shaped by external practices and individuals, via four primary discursive mediating processes: Alienation, Translation, Recontextualisation, and Absorption. In combination, these four movements illuminate both the path and effect of biotechnology mediation at multiple levels of social organisation and in particular contexts. It is precisely because of biotechnology's potentially wide reaching influence, and relative political and policy weighting in multiple countries at present, that we should seek to identify, comprehend, and evaluate its effects in contexts beyond those in which it is officially purported to operate or function. Before we can evaluate the socio-ethical, and socio-political effects of biotechnology we need to be able to 'see' how it operates across multiple contexts. This framework provides an analytical and systematic *lens* through which to view and evaluate the breadth and depth of macro level social, ethical, and political shifts that are associated with the inculcation of new technologies in general and biotechnologies in particular.

UNDERSTANDING BIOTECHNOLOGY AS MEDIATION

Understanding biotechnology as mediation requires us to step away from what we immediately think of as "media" for example television, radio, or print. Instead of looking at these established media *forms* or *genres*, I am seeking to explore contemporary biotechnology practice as a series of processes of mediation. In Silverstone's words, mediation involves

Naomi Sunderland et al. (eds.), Towards Humane Technologies: Biotechnology, New Media and Ethics, 187–205.

...the movement of meaning from one text to another, from one discourse to another, from one event to another. It involves the constant transformation of meanings, both large scale and small, significant and insignificant, as media texts and texts about media circulate in writing, in speech and audiovisual forms, and as we, individually and collectively, directly and indirectly, contribute to their production (1999, p. 13).

The phrase 'biotechnology as mediation' in this context refers to the multiple processes, contexts, and texts of mediation associated with biotechnology. The process of mediation and movement of biological and discursive resources for meaning making is, as Silverstone points out, 'fundamentally political or perhaps, more strictly, politically economic' (1999, p. 4).

Contemporary definitions of biotechnology esteem and embody mediation – usually by way of translational research, commercialisation, product development, or application. Sample definitions are included below.

'[Biotechnology is] The application of scientific techniques that use living organisms, or substances from those organisms, to make or modify products, improve plants and animals, or to develop micro-organisms for specific uses' (US Office of Technology Assessment).

'The use of biological systems - living things - to make or change products. It has been used for centuries in traditional activities like baking bread and making cheese' (CSIRO).

'Development of products by a biological process. Production may be carried out by using intact organisms, such as yeasts and bacteria, or by using natural substances (e.g. enzymes) from organisms' (International Industrial Biotechnology Association).

'Biotechnology is a very broad term referring to any practical or commercial use of living organisms, such as using yeast to make beer or bread' (Washington Biotechnology Action Council).

These typical definitions of biotechnology describe a given organism or biological process as a means to producing or acquiring a given technology-based use, outcome or product.

Within biotechnology's orientation toward application or product development, there are also a number of broader, historically salient mediating forces at hand. My understanding of the political and economic orientations of biotechnology as media is particularly influenced by Marcuse's (1964) analysis of the ways that modern societies can work to dilute and devalue any form of 'antagonistic' or 'subversive' public opinion (p. 9). Marcuse argues that technological innovation in the form of media in particular has allowed antagonistic/subversive content to be recontextualised into the "everyday" operations of the productive apparatus. The point is, that in modern biotechnological processes, something so wondrous as the foundations of life somehow are translated into the form of a product to which only a select few have access.

We take DNA, for example, from a hitherto 'secret' place, and move it progressively toward something that is part of our everyday existence such as vaccines, treatments, products, services, and so on. But not only is DNA moved, it is also changed, altered, politicised so as to *fit with* existing social trajectories or demands, and the paths of mediation on offer. Scientific discovery in effect loses its initial meaning and value by being diluted and subsumed under the commodity logic into which it is being recontextualised. Perhaps more correctly, the great scientific "discovery" has its meaning and value *reinterpreted* by different human agents, its scientific potential evacuated and replaced with something else that is relevant to its new context for example, the 'price system' (Innis, 1942).

This framework for viewing biotechnology as media is based on a specific understanding of biotechnology as a socio-political and socio-ethical practice rather than a simply technical one. While it may be a given understanding for some, this is important for several reasons. First, framing biotechnology as a social practice sets a theoretical foundation and starting point for analysing how mediating processes in biotechnology can and do affect human understandings of themselves and others in both implicit and explicit ways. Humans are embedded in multiple social practices, contexts, and ecologies at any given time. We learn about ourselves and others through our being in the world. Likewise, when we learn to conduct a specific practice or profession we are learning the ways of seeing, being, acting, and describing that characterise that practice and that have been handed down from generations before. Biotechnologists, like any other professional group, are subject to specific socialising influences that arise from both within and beyond the practice. The traditions that shape practice in biotechnology can come from both internal and external sources. Second, adopting a social practice view of biotechnology emphasises the fact that biotechnology practice has a history and a future that is both produced and reproduced by the persons that are involved both directly and indirectly in its creation and perpetuation in social life. Third, social practices are "separated" from their social environment, by discursively constituted boundaries that are produced and reproduced both within and beyond the practice itself (cf. Isaacs, 1998; Luhmann, 1995). A discursive boundary in this sense is quite simply one of meaning, description, and shared understanding.

The mediating movements of Alienation, Translation, Recontextualisation, and Absorption discussed in this chapter work beyond the social practice of biotechnology itself. As mentioned above, the boundaries between any social practice, other practices, institutions, professions, and "society" at large are discursively constituted and hence are *permeable*. The boundaries that separate the social practice of science from ethics or public policy, for example, are not made of electric fences. Rather, these boundaries are the products of shared and consistently reproduced ways of seeing, acting, being and describing that people share. It is not only the people within any given practice who create and sustain these boundaries. Because the boundaries are discursive and permeable, other social practices, institutions, individuals, and so on can also contribute to them, and to the practice itself. The influence that biotechnology has on the rest of society is not one-way. Other practices and agendas can intertwine with what might be seen to be the "core" purpose or "business" of biotechnological research.

THE FOUR MOVEMENTS OF BIOTECHNOLOGY AS MEDIA

This chapter seeks to establish a way of understanding exactly how a practice such as biotechnology can result in broad social and ecological change. I outline four primary mediating movements to explain how biotechnology operates as media in multiple social and ecological contexts beyond the laboratory. The four mediating movements discussed in this chapter are not intended to be mechanistic or linear. Rather, at any particular moment of mediation the four mediating processes can be identified in multiple contexts and practices, in different orders of progression, and even simultaneously. Indeed, alienation, translation, recontextualisation, and absorption are closely featured in most, if not all, biotechnology mediations. Each of the processes is an integral *dimension of* mediation in, and surrounding, the social practice of biotechnology. As the following sections illustrate, each of the four processes are in fact *required* if biotechnology is to develop according to the values, objectives, and purposes prescribed in state, national, and trans-national policy texts.

Alienation

Alienation has once again become a point of interest in discussions regarding intellectual property rights for biotechnology processes and products (cf. Andress and Nelkin, 1998; Flowers, 1998; Nelkin and Andrews, 1998; Thompson, 1995). However the main reason the term has re-emerged is not to discuss the alienation of human labour from human beings or the ultimate alienation of humans from nature, but as a precondition for commoditising bio-products (cf. Thompson, 1995). Thompson states that 'a good or the right to enjoy a good is "alienable" to the extent that it can be *dissociated from one owner of the good and transferred to another*' (p. 281, italics added). Rivalry on the other hand 'refers to the situation where the use or consumption of the good by one person diminishes the amount of good available for others' like, for example, a can of tomato soup (Thompson, 1995, p. 281). Lighting on public streets on the other hand is in most cases a non-rival good. Excludability refers to how easy or hard it is to *exclude others* from using a particular good (Thompson, 1995, p. 281). By consuming a can of soup, a person excludes the possibility of anyone else consuming it. On the other hand it would in most cases be very difficult to restrict others from accessing the light thrown off from street lamps in your neighbourhood.

The three conditions of alienability, rivalry, and excludability, according to Thompson, are the prerequisites for declaring something as property. Thompson's explanation is particularly effective:

> One person cannot listen to the song of a dove while someone else eats the same dove roasted, because these are rival uses of the dove. However, if the dove's song is alienated from the dove itself with a recording, previously rival uses become associated with separable goods, their rivalry diminishes, and the potential for hearing the dove's song while feasting on its flesh becomes possible. *Alienability of a good is thus a necessary condition for*

regarding it as exchangeable property. (Thompson, 1995, p. 281, italics added)

Thus, the process of *alienation* – as Thompson would describe it – in the context of this framework is where the biotechnology as a technological medium function is most concentrated: Humans use modern genetic technologies to *dissociate* biological materials from one 'owner' (plant, animal, human, or other living organism) or context to another 'owner' (in the form of "intellectual property") or context (for example, DNA shifted from the context of the body to a laboratory setting or computer database). Recombinant DNA techniques used in contemporary biotechnology are significant in that they

> ...represent means to alienate goods from previous patterns of ownership and exchange and to establish new rights of ownership and exchange. Although conventional plant or animal breeding was capable of introducing substantial changes in the traits or composition of individuals, it did not permit the alienation of those goods from representative individuals themselves. (Thompson, 1995, p. 282)

Although it is a radical movement, alienation in itself is not sufficient to explain the way that previously inalienable aspects of life are transferred into economic goods for exchange on the commercial market. Indeed, the significance of alienated genetic 'codes', cells, DNA and so on are, in their purely technical forms, accessible only to those discourse communities versed in the discourses associated with Recombinant DNA technologies and molecular genetics. For the alienated 'goods' to be taken beyond the laboratory context, they must undergo several interrelated processes of meaning production and reproduction, including translation, recontextualisation, and absorption.

Translation

> We cannot accurately conceive what it must have been like to be the first to compare the colour of the sea with the dark of wine or to see autumn in a man's face. Such figures are new mappings of the world, they reorganise our habitation in reality. (Steiner, 1975, p. 23)

Translation is the process of recasting a system of meaning in the form of another, often fundamentally different system of meaning. Translation is the most overt discursive function of the four media movements. Translation is inherently political and interpretive. Far from being a process concerned with opening access to new spaces, alienation can be seen as a process of translation and encoding (or codifying) rather than "decoding" the human genome; it is the translation and compacting of previously inalienable meanings and biological resources into alienable scientific discourse, applications, material, biological products, and so on (cf. for example DeCode Genetics' corporate logo "decoding the language and life", nd, np). In Steiner's words, '[t]he translator invades, extracts, and *brings home*' (1975, p. 298). Steiner's use of the word "extracts" is particularly pertinent in that it emphasises the selective and interpretive nature of translation, and its potentially minimising filtering effects on the previously inalienable, merely potential meaning or resource. Like Steiner, Silverstone argues that in 'translation we enter a text and claim ownership of its meaning' (1999, p. 15). Translation is 'a move which involves both meaning and value. While it might seem at times a largely technical or pragmatic activity, translation is in fact 'both an aesthetic and an ethical activity' (Silverstone, 1999, p. 15).

> Translation is process in which *meanings are produced*, meanings that cross boundaries, both spatial and temporal. To enquire into that process is to enquire into the instabilities and flux of meanings and into their transformations, but also into *the politics of their fixings.* (Silverstone, 1999, p. 16, italics added)

Because it is a process of *transforming* and, according to Steiner (1975), *fixing* meaning, translation also involves limiting potential "audiences" (or discourse communities) who may access these newly translated meanings. For example, the human genome directory is expressed in a series of codes that use four letters of the alphabet. The discursive boundaries around the social practice of biotechnology are very clear. Thacker's (2000) work on the translation of the body as data is perhaps most helpful in deciphering what is actually happening in this process (see Thacker chapter 'Data made flesh' in this volume). Here is an example from an article titled 'A crystallographic map of the transition from B-DNA to A-DNA' which uses the four letter coding system for DNA in written form:

> The transition between B- and A-DNA was first observed nearly 50 years ago. We have now mapped this transformation through a set of single-crystal structures of the sequence $d(GGCGCC)_2$, with various intermediates being trapped by methylating or brominating the cytosine bases. The resulting pathway progresses through 13 conformational steps, with a composite structure that pairs A-nucleotides with complementary B-nucleotides serving as a distinct transition intermediate. The details of each step in the conversion of B- to A-DNA are thus revealed at the atomic level, placing intermediates for this and other sequences in the context of a common pathway (Vargason, Henderson, and Shing Ho, 2001, p. 7265).

Granted, the article by Vargason et al. is not intended for general audiences –and thus is not written in a language that general audiences will understand –yet it is obvious to all that the sequencing "$d(GGCGCC)_2$" is an exclusive *encoding* translation of the human genome. As an earlier reviewer of this chapter notes, those who are able to read this language may take several years to become fluent in this language of DNA in much the same way that one might take several years to become fluent in Greek or Arabic. Vargason et al.'s use of terminologies and phrases such as 'A-nucleotides', 'distinct transition intermediate', and 'common pathway' is similarly mystifying to an outsider-audience, despite the fact that they still employ common English vocabulary (i.e. commonly accessible words such as distinct, transition, intermediate, common, pathway, etc).

The direct translation process from the previously inalienable "language of life" into technocratic scientific discourses and modes of representation is apparently separate from any contemplation over the ethical and moral aspects of the technologies. As we move away from DNA, we see further translations and recontextualisations at play, from scientific discourse to the discourses of economic and entrepreneurial enterprise, bureaucratic administration, marketing, promotion, commoditisation, financial futures, regulation, health, therapy, conception, motherhood, humanity, and so on.

In the 1998 *Hastings Centre Report*, Nelkin and Andrews (1998) make the following translations into the technocratic discourse of market economics (see Vogel chapter in this volume):

These expanding *markets* have increased the *value* of human tissue, and institutions with ready access to tissue find they possess a capital resource. Access to stored tissue samples is sometimes included in collaborative agreements between hospitals and biotechnology firms. (Nelkin and Andrews, 1998, p. 30, italics added)

Note here that the agents are different from those in the example from Vargason et al: here examples of agents include markets; collaborative agreements; hospitals; and biotechnology firms. Obviously, the biotechnological context Nelkin and Andrews are talking about is a different one from Vargason et al, and at a different stage of mediation and recontextualisation. Note particularly that 'expanding markets' are the agents that have imbued human tissue with 'increased ... value'.

All processes of mediation, just like technologies themselves, are imbued with inscribed value judgements as to which biological resources are desirable and important in any setting – not merely in the context of economic exchange (cf. Martin, 2000). In its crudest form, commercial "viability" becomes a litmus test for which products will become readily available resources for making meaning within the broader discourses to which biotechnology research and commercialisation are shifted. It is implicitly accepted, therefore, that 'free market acceptance of a good or technology [is] equivalent to an ethical endorsement' (Thompson, 1995, p. 276).

Even if members of the general community cannot access the technocratic discourses surrounding the translation and absorption processes, they are able to draw their own meanings from the everyday productive manifestations of these processes or specific programs to influence social understanding (and usually acceptance) of biotechnology practices and products. Non-scientists and non-economists can draw on the commodified version of biotechnology as media simply by buying it, selling it, using it, being aware of it, and so on. Hence *the process of translation is enacted once again at the point of capital exchange and consumption.*

General understandings of biotechnology research and development are conveyed through the everyday sale of commercial goods and services, as well as through wider media and political discourses and rhetoric surrounding biotech-nology processes themselves. Biotechnology products and services, as manifest-ations of biotechnology as media, are literally *absorbed* via consumption into the everyday lives of members of the public. Biotechnology techniques and technologies themselves are in many cases the product of a long process of alienation, translation, recontextualisation, and absorption with imbued values, motivations, and judgements as to their viability, worthiness and so on.

The already abundant range of bio-products and services circulating in global pharmaceutical and "life" markets indicates the extent to which previously inalienable or incomprehensible aspects of human life have already been absorbed in the everyday productive apparatus:

> In recent years, biotechnology techniques have transformed a variety of human body tissue into valuable and marketable research materials and clinical products... the catalogue from the American Tissue Culture Catalogue lists thousands of people's cell lines that are available for sale. Body tissue also has commercial value beyond the medical and research contexts. Placenta is used to enrich shampoos, cosmetics, and skin care products...There is also a market for services to collect and store one's tissue outside the body. People can pay to store blood prior to surgery or embryos in the course of in vitro fertilization...There are about fifty private DNA testing centers in the United States, hundreds of university laboratories undertaking DNA research, and over 1,000 biotechnology companies developing commercial products from bodily materials. (Nelkin and Andrews, 1998, p. 30)

At each stage of translation, recontextualisation, and absorption, the discourses, and physical bio-products themselves are *produced as a result of*, certain technocratic practices and orientations (cf. Martin, 2000). Attributions of value, desirability, worthiness, ethicality and so on are specific to the particular discourse communities and social contexts. For example, where the successful cloning of a human being might be seen as a major scientific breakthrough for the scientific community and those who will directly benefit from the technology, cloning a human may be seen as an immoral practice fraught with danger and fear for other members of the general community. Individuals involved in discourse communities concerned with ethics or public policy may also *attribute* the scientific "breakthrough" with different meanings, consequences and so on. Obviously, the social embeddedness of all social practices, and the persons that constitute them, ensure that they do not operate in complete isolation from one another. But, when the practices and processes associated with biotechnology enter distinctly different social systems and settings (such as the system of commercial exchange, regulation, or politics) their meanings, significance, perceived value, desirability, degree of familiarity identifiably change.

All uses of the term "translate" in biotechnology policy documents are consistent with the use of the term in relation to mediation. There is an acknowledgement that a movement from one context to another both precipitates and requires translation or fundamental shifts in meaning systems. The following excerts provide examples of translations that are involved in contemporary biotechnology scenarios.

BioPLATFORM will: - extend the foundation for translating excellence in research into economic benefits' (NSW Govt Biofirst 2001 Strategy);

'Australia has strengths in scientific discovery, which are not currently being translated into exploitable intellectual property' (Victorian Strategic Plan 2000);

'It is essential to have the capacity to translate knowledge into new products, processes and services, that in turn will generate benefits to society, skilled jobs and prosperity' (A Strategy for Europe);

'Industry...has a key role in translating our research base into products, services and wealth' (Victorian Strategic Plan 2000);

California-based Genentech has made remarkable progress in translating genetic information into tangible, practical information to change drug development (Black Art Industry magazine article).

The orientation of translations portrayed above is different from the orientations that feature in the official rationalisation of biotechnology: that it is intended to improve our standard of living, health and well being for *all*.

Recontextualisation

The process of translating an entire, and until recently inaccessible, aspect of life into commodifiable products and services is also highly dependent on processes of *recontextualisation* (cf. Bernstein, 1990; Iedema, 1997a). Recontextualisation is the process by which discourses are encapsulated in 'increasingly durable materialities' as a direct result of their translation and entry into new social systems and contexts (Iedema, 1997a). The consequence of recontextualisation then is that these 'increasingly durable materialities' – such as a technology or a product – are seen to encapsulate the discourses that have shaped their being and becoming: They are discourse materialities.

Sarangi (1998) extends on Iedema's work to emphasise that recontextualisation is necessarily coupled with processes of decontextualisation and entextualisation. He observes that

> ...putting something into context (contextualizing it), putting something out of context (decontextualizing it), and putting something into a different context (recontextualising) are both everyday and scientific activities...In between decontextualization and recontextualization, Bauman and Briggs (1990) suggest, there is a process of *'entextualization'* in narrative performance: an event is entextualized into a discourse with a controllable set of truth-values...Recontextualization is thus not 'representation', but 're-presentation' or re-production' which implies creativity (Sarangi, 1998, pp. 306-7, italics added).

Recontextualisation, then, requires concomitant processes of decontextualisation, transformation, and *entextualisation*[xxii]. For example, when pieces of foreskin tissue or placenta are alienated from their origin body-context and re-contextualised into other contexts such as the laboratory, clinic, or hospital, the original and typically tacit value and meaning of "foreskin" or "placenta" are replaced by other overt and functional meanings, use values, and exchange values by agents in these new contexts. A point to note also is that the range of persons who have power to entextualise the foreskin and placenta with meaning and value, and to use it for specific purposes in research or treatment contexts, are vastly different from the original and personal 'owner' of the tissue in its body-context.

Both Iedema and Sarangi see the creation of written texts as a significant part of the recontextualisation process. Texts in particular are significant indicators of which meanings and values get 'left behind' and which are foregrounded at various stages of recontextualisation (cf. Lojek, 1994, p. 84; Mehan, 1993; Sarangi, 1998, p. 308).

The following excerpt from the UN Human Development Report 2001 provides a detailed example of the role of texts in recontextualisation. recontextualising process from the corpus. The excerpt details the steps by which the UK Government's Technology Foresight Program moves from its initial conception to being implemented in various contexts, in various forms, by various agents. I have underlined the different contexts and genres that the 'UK Technology foresight program' moves through toward its 'applications' and 'outcomes'. The contexts range from particular institutional genres, such as 'the steering committee' to more abstract discursive formations, such as 'four themes'.

The UK technology foresight programme, announced in 1993, is forging a closer partnership between scientists and industrialists to guide publicly financed science and technology activity [hybridity]. More market oriented and less science driven than similar efforts elsewhere, the programme has had three phases. First it set up 15 panels of experts on the markets and technologies of interest to the country, each chaired by a senior industrialist. Each panel was charged with developing future scenarios for its area of focus, identifying key trends and suggesting ways to respond. In 1995 the panels reported to a steering group, which synthesized the main findings and identified national priorities. Next the steering group produced a report distilling its recommendations under six themes: social trends and impacts of new technologies; communications and computing; genes and new organisms, processes and products; new materials, synthesis and processing; precision and control in management, automation and process engineering; and environmental issues. The steering group assigned priorities to three categories: key technology areas, where further work was vital; intermediate areas, where efforts needed to be strengthened; and emerging areas, where work could be considered if market opportunities were promising and world-class capabilities could be developed. Now the recommendations from the exercise are being implemented. For example, research in the four priority areas- nanotechnology, mobile wireless communications, biomaterials and sustainable energy-is being supported through a research award scheme. Another example is its application in Scotland. Scottish Enterprise hosts the Scottish foresight coordinator, who focuses on promoting foresight as a tool for business to think about and respond to future change in a structured way. The coordinator works with a wide range of public, private and academic actors. While a key goal is to help individual companies better manage change, this is being achieved by channelling efforts through a range of trusted business intermediaries-industry bodies, networks and local delivery organizations -that have a sustainable influence on company activities. All panels and task forces address two underpinning themes: sustainable development and education, skills and training. (UN Human Development Report)

A point to note regarding the processes of recontextualisation illustrated above is that the point of motivation, concern, exchange, interest, or desire associated with the Technology Foresight Program underwent multiple transformations that were beyond the reach of the original program authors. The reader will note that recontextualisation in this example entails shifts not only in text types and genres (from reports to meetings, to strategies, to taskforces, to training), but also shifts between contexts and agents: who is involved, which institutions, which spaces, which practices.

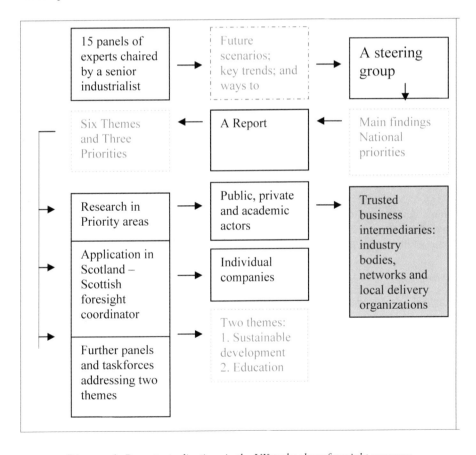

Diagram 1: Recontextualisations in the UK technology foresight program

I have myself 'translated' the above excerpt into the diagram below in order to illustrate these movements more clearly. The boxes that feature dotted borders are intended to convey discourse formations such as "themes" or "priorities" rather than particular social contexts or texts which are presented in solid boxes. Interestingly, it is the themes or priorities that are the currency of recontextualisation in this example: they are the links in a chain between one context and another. The themes and priorities themselves are heavily condensed summaries of all of the discussions and interests that have emerged in prior contexts. They are nominalisations (i.e. verbs/actions converted to nouns/things).

A final point to note in regard to the movement of recontextualisation is that recontextualised abstractions, when absorbed in objectified material culture, such as the various bio-products now on offer, themselves become 'a component in the process of communication' (Streeck, 1996, p. 366, in Iedema, 1997a): that is, a consumable *resource* for meaning making and evaluation. As intimated above, many members of the general public – i.e. those who are not members of the technocratic genetics discourse community – will, to a certain extent, draw on the objectified material culture as resources for meaning making. The material manifestations of biotechnology themselves are imbued with, and indeed ontologically produced and defined by, a set of technocratic (economic and scientific) values, judgements, and evaluations.

But, as discussed in the Introduction, the durable materialities of biotechnology recontextualisation are not merely 'products' such as a particular diagnostic kit or vaccine, they are also living plants, animals, humans, and other organisms such as bacteria. These outcomes of biotechnology become a component in the process of communication. We may come to understand and evaluate GM foods, for example, through their availability for sale in supermarkets. We may have a friend who aborted a child because of the results of a genetic test taken during the early stages of pregnancy. We may have a nephew who was conceived through reproductive technology. These persons are within our families, our friends, and colleagues. These persons, whose lives have been directly manipulated by biotechnological mediation, may choose to contribute their own 'lived experience' evaluations of these technologies to public discourse.

Absorption

Technological innovation provides a means by which humans can move previously non-routine aspects of human cultural expression and life to into "everyday" mediated contexts. Marcuse's (1964) description of the way that abstract cultural expressions and antagonistic or subversive cultural content are depleted and transformed (homogenised so as to fit within the media form) via mediation is particularly instructive here. Marcuse (1964, p. 61) argues that mediation affects not only how things appear, but also where they appear, and in what form (for example, the salon, the concert hall, the theatre, the market) . All of these variables, he argues, impact upon the perceived social significance, meaning, and political potential of both the media form and its "content" (Marcuse, 1964, p. 62).

Marcuse contends that the antagonistic or subversive potential of content – that is, the potential to be other than what *is* and to effect consciousness of, or desire for, something other than the current path of mediation – is depleted by 'the absorbent power' and more or less "everyday" status of a particular media form. A particular media form, for instance, cannot help but filter and shape the content that is passed through it because of its own limitations, transmission channels, intended uses, and so on. The *context within which* an "audience" interprets content (for example, a theatre, concert hall, laboratory, supermarket), in conjunction with the filtering effects of a media form itself, is also significant to the range of possible meanings that are attributed to that content by the audience or "consumer".

> The absorbent power of society depletes the artistic dimension by assimilating its antagonistic contents. In the realm of culture, the new totalitarianism manifests itself precisely in a harmonizing pluralism, where the most contradictory works and truths peacefully coexist in indifference…Whether ritualised or not, art contains the rationality of negation. In its advanced positions, it is the Great Refusal – *the protest against that which is*. The modes in which man [sic] and things are made to appear, to sing and sound and speak, are modes of refuting, breaking, and recreating their factual existence. But these modes of negation pay tribute to the antagonistic society to which they are linked. Separated from the sphere of labour where society reproduces itself and its misery, the world of art which they create remains, with all its truth, a privilege and an illusion…The salon, the concert, opera, theatre are designed to create and invoke another dimension of reality. Their attendance requires festive-like preparation; *they cut off and transcend everyday experience* (1964, pp. 61-63, italics added).

The key aspect of absorption, as distinct from the other movements of mediation, is that it deals specifically with these processes of rendering new technologies familiar, invisible, and part of the "everyday". All of the movements of mediation discussed above are, however, intimate in this inherently political process of absorption. As the following diagram indicates, all of the movements of mediation are intimately involved in moving biotechnologies, and the living products they engender, into the everyday lives of citizens and ecologies. Absorption in biotechnology requires a movement from inalienability to commoditisation; from abstraction to absorption; and from spaces and times where the technology or product is new and contested to spaces and times where is nothing more than an everyday, acceptable product or service, and familiar (see figure below).

An integral part of the absorption movement is, then, the process of rendering a new technology desirable, acceptable, and familiar. This movement may have to occur in an atmosphere of considerable opposition, as has been the case with some areas of biotechnology including GM foods, cloning, and stem cell research. Among other things, dissent disrupts the invisibility of a new technology and focuses, rather than deflects, critical consciousness.

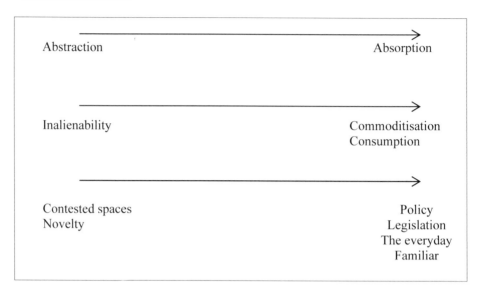

Figure 2: Abstraction, absorption, alienation, and commoditisation

The rhetorical imperative is paramount in some areas of technological absorption because the hopes that surround technological revolutions are, characteristically, not the hopes of the many – at least in their first incarnation. Rather, the nature of technological 'advance' is such that new technologies have to be "introduced" to social systems and be rendered familiar, acceptable, and desirable through strategies of influence and persuasion. In mediation, even technologies that have been hotly contested can become part of the accepted and familiar everyday by becoming familiar, available, and gradually indispensable. As such, the most crucial role of the introducers of new technology is to make others desire the new technology, and the outcomes it may accrue. In high technology industries, the most important people to convince are arguably the funding agencies which, in current climes, equals industry representatives and politicians.

New media forms and products find their way into the everyday as material objects and processes and even as further abstractions of capital and value, for example, spin off companies, stocks and bonds, currency, and "futures". These products, technologies, and their intangible abstractions become part of the everyday through rhetorico-political processes of routinisation and naturalisation. The need to 'introduce' technologies to a society, and to secure their 'place' in that society, is manifest in the imperatives of industrialists and academics who have become increasingly concerned with processes of 'technological diffusion; 'innovation infusion'; 'early adoption'; 'early adopters'; 'critical mass'; 'media saturation'; and so on (cf. Green, 2002; Hauben and Hauben, 1997; Takacs and Freiden, 1998).

THE STATE OF MEDIATION

The reaction of the state power upon economic development can be one of three kinds: it can run in the same direction, and then development is more rapid; it can oppose the line of development, in which case nowadays state power in every great nation will go to pieces in the long run; or it can cut off the economic development from certain paths, and impose on it certain others. (Frederick Engels, Letter to Conrad Schmidt, Oct. 27 1890, in Marx and Engels, 1947, p. 4)

The current pathways of biotechnology mediation have been officially sanctioned and promoted by governments across the globe. In Engels' terms above it seems that governments have chosen the "run with it" option. Yet, the rhetorical enhancement of technologies and common declaration of technological revolution that is evident in much public policy on biotechnology is in no way insignificant or rare. One may even argue that the declaration of technological revolution in these policy documents has become little more than a rhetorical device designed to inscribe new technologies with unquestioned overtones of social betterment and the improvement of human life. The most pervasive and uncritical claim in the discourses of technological revolution is that a technologically 'revolutionised' society is one that is more 'advanced' than others. A technologically revolutionised country is, apparently, forward looking, bold, innovative, modern, cutting edge, and leading the way. Moreover, to join the 'industrialised' world, developing countries are told that they must adopt the latest technologies, and perhaps more significantly, the dogma that accompanies their inculcation.

Like all technological 'revolutions' of our history, biotechnology is defined by profound hopes for the future. Whether these hopes match the expectations of those who initiated them or not, the outcomes of any technological revolution are of great consequence. This is because technological revolution hinges on transformation: Technological revolution happens when some thing, society, process, practice, or one, is transformed. But a revolution does not necessarily mean that the dominant modes of *valuing*, worldviews, or ideologies are transformed. The most prolific transformations, it seems, are in fact rendered upon what can be transformed: what is the object of technology. Technological revolution is deemed to be revolutionary to the extent that it opens up new or expanded spaces of physical, biological, and social life for exploitation according to the dominant means and imperative of the system from which it is borne and into which it is [re]introduced. Technological revolution does not, historically, stop exploitation from occurring but, rather, simply increases the range of natural human and non human life forms that are subject to productive exploitation.

Once a new technology has been factored significantly into a nation's policy mix it becomes authorised, rather than aberrant and strategic rather than abstract. The new technological 'innovation' is recontextualised as the stuff of enlightened public policy. As Feenberg (1999) posits, this focus on technology and scientific revolution as a precipitant of human progress is a characteristic feature of modern western societies:

There is, however, another fateful path by which technology enters the larger conversation of modernity: the historicizing trend in the emerging biological and social sciences of the late 18th and 19th centuries. This trend was firmly rooted in the idea of progress, which found its surest guarantee in the promise of technology. By the end of the 19th century, under the influence of Marx and Darwin, progressivism had become technological determinism. Following the then common interpretation of these materialist masters, technical progress was believed to ground humanity's advance toward freedom and happiness (Feenberg, 1999, pp. 1-2).

Once advancements in a particular area of technological development becomes policy, the state moves henceforth as the primary medium of technological-ideological diffusion: that is to say as the primary force of mediation. The 'new' – rather than 'aberrant' – technology is literally 'frameshifted' to the level of national and international policy and consciousness (Waller, 2001). Through the medium of the state and productive apparatus, both the material and non material aspects of the new dogma enter schools, universities, public institutions, public spaces, homes, ecosystems, and bodies. The substantive basis of these movements from idea to the everyday involves the 'interlocking apparatus of scientific research, technological innovation, and industrial mass production' (Leiss, 1994, p. xii). Such paths of officially sanctioned mediation are difficult to contest precisely because official rationales for mediation become ensconced in representations of the (future) good life. The result is that anyone who argues against it is often rhetorically pitted against the 'future wellbeing of all of humanity', or the economic growth and prosperity of their nation and their children.

These diffusion, routinisation, and naturalisation strategies are predominantly carried out via existing social media including the State, mass communication media, and markets of exchange. As new technologies emerge, the institutions of governance, law, and even ethics are invoked to regulate and patrol the development and diffusion of the technology and to advocate in the interests of public 'safety'. Programs also emerge, usually post facto the initial surge of economic activity and division of property rights, in an attempt to ensure equality of access to the materialisations and capital abstractions of the new technology. An important part of the routinisation of new technologies is the phenomenon whereby 'access' to the material outcomes of a new technology becomes a basic human 'right': a measure of human development and the good life. When this occurs, it is assumed that all persons should have access to these technologies because they are unquestionably good, beneficial, and desirable. A person who does not have "access" to the new technology can be literally excluded from accessing some very basic social services which now depend upon that technology – the use of electronic banking over face to face and services delivered via the Internet is a classic example that has led to declarations regarding 'the digital divide' etcetera.

The most prominent example of this in the context of biotechnology is in relation to the lack of access to HIV/AIDS drugs in sub Saharan Africa. What the UN (2001, cwn. 65,919) has termed "Poor People's Technology" programs (publicly funded technology) have been introduced to ensure that 'disadvantaged' communities in 'developing' nations have the same opportunities to access drugs and vaccines as rich people in western industrialised countries (United Nations, 2001, cwn.72,647). The function of not for profit programs and social policies run out of the United Nations and other social and environmental justice bodies are to attend to persons or ecosystems who/that may be left out of the dominant modus operandi of the large pharmaceutical companies. What these groups are responding to is the fact that some people are literally *rendered* vulnerable by the current paths of mediation and development in biotechnology. As the authors of the UN Report state, the role of the UN's Poor People's Technology programs is to fill in gaps that are created by 'market failure'.

CONCLUSION

Biotechnology is defined in public policy and organisational discourses as a branch of science practice that is primarily concerned with commercial, material, and product outcomes. This is despite some scientists' claims that the primary function of the practice is to contribute to the stock of human knowledge and understanding, environmental benefit, alleviation of hunger, or the provision of new drugs and pharmaceuticals to those who suffer. Through convergent technologies of biology and information humans can use biotechnology to increase the range of human and non human living organisms that fall under the commodity logic of contemporary capitalism. In this biotechnology 'revolution' a broader range of humans, animals, and plants are rendered – in a more thorough way – both materials and sites of capital production. All of the primary mediations (alienation, translation, re-contextualisation, and absorption) that form biotechnology research and commer-cialisation make the distance between the initiators of scientific discoveries, and those who ultimately are affected by or consumers of the technology, or product further and further apart. With each new context that is identified in biotechnology and its related practices, new actors are also either identified or implied. So too, with every change of context do we see new representations and offspring of biotechnology emerge

NOTES

xxii In later sections of the chapter I use the one term 'recontextualisation' to refer to the related processes of decontextualisation, recontextualisation, and entextualisation.

REFERENCES

Andress, L., & Nelkin, D. (1998, January 3). Whose body is it anyway? Disputes over body tissue in a biotechnology age. *The Lancet, 351*(9095), 53–57.

Bernstein, B. (1990). *The structuring of pedagogic discourse. Class codes and control* (Vol. 4). London: Routledge.

Commission of the European Communities (CEC). (2002). *Life sciences and biotechnology - A strategy for Europe*. CEC.

CSIRO. (2000a). *Biotechnology: What is it?* CSIRO on-line resource.

CSIRO. (2000b). *Gene technology: How is it done?* CSIRO on-line resource.

Feenberg, A. (1999). *Questioning technology*. London: Routledge.

Flowers, E. B. (1998, November). The ethics and economics of patenting the human genome. *Journal of Business Ethics, 17*(15), 1737–1745.

Green, L. (2002). *Technoculture: From alphabet to cybersex*. St Leonards, NSW: Allen & Unwin.

Hauben, M., & Hauben, R. (1997). *Netizens: On the history and impact of Usenet and the Internet*. Los Alamitos, CA: IEEE Computer Society Press.

Iedema, R. (1997a, July 2–6). Bureaucratic planning and resemiotisation. In E. Ventola (Ed.), *Proceedings of the Ninth Euro-International Systemic Functional Workshop*. Halle-Wittenberg, Germany.

Innis, H. A. (1942). The newspaper in economic development. *Journal of Economic History, 2*(December), 1–33.

International Industrial Biotechnology Association.

Isaacs, P. (1998). *Social practices, medicine and the nature of medical ethics*. Paper presented at the Society for Health and Human Values Spring Regional meeting. Youngston State University, Youngston, Ohio.

Leiss, W. (1994). *The domination of nature*. New York: Beacon.

Lojek, H. (1994, Spring). Brian Friel's plays and George Steiner's linguistics: Translating the Irish. *Contemporary Literature, 35*(1), 83–93.

Luhmann, N. (1995). *Social systems* (J. Bednarz & D. Baecker, English trans.). Stanford, CA: Stanford University Press.

Marcuse, H. (1964/1968). *One dimensional man*. London: Routledge.

Martin, J. R. (2000). Beyond exchange: Appraisal systems in English. In S. Hunston, & G. Thompson (Eds.), *Evaluation in text: Authorial stance and the construction of discourse*. Oxford, UK: Oxford University Press.

Marx, K., & Engels, F. (1947). *Literature and art*. New York: International Publishers.

Mehan, H. (1993). Beneath the skin and between the ears: A case study in the politics of representation. In S. Chaiklin & J. Lave (Eds.), *Understanding practice: Perspectives on activity and context* (pp. 241–268). Cambridge: Cambridge University Press.

Nelkin, D., & Andrews, L. (1998, September–October). Homo economicus: Commercialization of body tissue in the age of biotechnology. In *The Hastings Center Report*. New York: The Hastings Center.

New South Wales Government. (2001). *Biofirst: NSW biotechnology strategy*. NSW Government.

Sarangi, S. (1998) Rethinking recontextualisation in professional discourse studies: An epilogue. *Text, 18*(2), 301–318.

Silverstone, R. (1999). *Why study the media?* Thousand Oaks, CA: Sage Publications.

Steiner, G. (1975). *After Babel*. New York and London: Oxford University Press.

Takacs, S. J., & Freiden, J. B. (1998, Summer). Changes on the electronic frontier: Growth and opportunity on the World Wide Web. *Journal of Marketing Theory and Practice, 6*(3), 24–37.

Thacker, E. (2000, June 21–25). *Data made flesh: Biotechnology and the discourse of the posthuman*. Paper presented at the Crossroads in Cultural Studies 2000 Conference, University of Birmingham.

Thompson. (1995, April). Conceptions of property and the biotechnology debate. *Bioscience, 45*(4), 275–289.

United Nations. (2001). *Human development report 2001*. United Nations.

Vargason, J. M., Henderson, K., & Shing Ho, P. (2001). A crystallographic map of the transition from B-DNA to A-DNA. *Proceedings of the National Academy of Science (PNAS), 98*(13), 7265–7270.

Victorian Government. (2002). *Biotechnology strategic development plan for Victoria*. Victorian Government.

Waller, J. C. (2001). Ideas of heredity, reproduction and eugenics in Britain, 1800–1875. *Studies in History and Philosophy of Biological and Biomedical Sciences, 32*(3), 457–489.

Washington Biotechnology Action Council.

Wildman, W. J. (1999, May). The use and abuse of biotechnology: A modified natural-law approach. *American Journal of Theology and Philosophy, 20*(2), 165–179.

Naomi Sunderland
Griffith Abilities Research Program
Griffith University

Printed in the United States
by Baker & Taylor Publisher Services